Catalytic Processes for the Production of Automotive Gasoline

This book serves as the most comprehensive and advanced resource available on catalytic processes to produce automotive gasoline. It discusses enhancers and emerging technologies such as the obtention of gasoline from biomass, gasoline from carbon dioxide, and synthetic gasoline: MTG and Fischer–Tropsch. It concludes with a general outlook on the future of gasoline production and a discussion of electric vehicles.

A valuable reference for researchers and advanced students, this text also provides a strong basis for professional engineers, scientists, and entrepreneurs in the oil, gas, and energy industry.

In general, this book:

- Explains all aspects of gasoline production, including principles and recent developments
- Discusses the use of GHGs such as CO_2 to produce gasoline/fuels
- Discusses the different additives to enhance the gasoline performance and stability
- Covers oil- and non-oil-based catalytic processes
- Contrasts the new electromobility technologies and contemplates the future of gasoline use trends

Catalytic Processes for the Production of Automotive Gasoline

Edited by
Heriberto Díaz Velázquez

CRC Press
Taylor & Francis Group
Boca Raton London New York

CRC Press is an imprint of the
Taylor & Francis Group, an **informa** business

Designed cover image: artjazz/shutterstock and i viewfinder/shutterstock

First edition published 2025
by CRC Press
2385 NW Executive Center Drive, Suite 320, Boca Raton FL 33431

and by CRC Press
4 Park Square, Milton Park, Abingdon, Oxon, OX14 4RN

CRC Press is an imprint of Taylor & Francis Group, LLC

© 2025 selection and editorial matter, Heriberto Diaz Velazquez; individual chapters, the contributors

Reasonable efforts have been made to publish reliable data and information, but the author and publisher cannot assume responsibility for the validity of all materials or the consequences of their use. The authors and publishers have attempted to trace the copyright holders of all material reproduced in this publication and apologize to copyright holders if permission to publish in this form has not been obtained. If any copyright material has not been acknowledged please write and let us know so we may rectify in any future reprint.

With the exception of Chapter 5, no part of this book may be reprinted or reproduced or utilized in any form or by any electronic, mechanical, or other means, now known or hereafter invented, including photocopying and recording, or in any information storage or retrieval system, without permission in writing from the publishers.

Chapter 5 of this book is freely available as a downloadable Open Access PDF at http://www.taylorfrancis. com under a Creative Commons Attribution-Non Commercial-No Derivatives (CC-BY-NC-ND) 4.0 license.

Any third party material in this book is not included in the OA Creative Commons license, unless indicated otherwise in a credit line to the material. Please direct any permissions enquiries to the original rightsholder.

The funding bodies for the present work are Consejo Nacional de Humanidades, Ciencia y Tecnología (CONAHCyT) (project number CF-191973), and Mexican Petroleum Institute (project number Y.62011).

For permission to photocopy or use material electronically from this work, access www.copyright.com or contact the Copyright Clearance Center, Inc. (CCC), 222 Rosewood Drive, Danvers, MA 01923, 978-750-8400. For works that are not available on CCC please contact mpkbookspermissions@tandf.co.uk

Trademark notice: Product or corporate names may be trademarks or registered trademarks and are used only for identification and explanation without intent to infringe.

ISBN: 978-1-032-84435-0 (hbk)
ISBN: 978-1-032-89071-5 (pbk)
ISBN: 978-1-003-51728-3 (ebk)

DOI: 10.1201/9781003517283

Typeset in Times
by SPi Technologies India Pvt Ltd (Straive)

For Product Safety Concerns and Information please contact our
EU representative: GPSR@taylorandfrancis.com
Taylor & Francis Verlag GmbH,
Kaufingerstraße 24,
80331 München, Germany.

Contents

Author

Heriberto Díaz Velázquez was born in Tenango del Valle, Mexico. He earned a bachelor's degree in chemistry at the Autonomous University of the State of Mexico, a master's degree in materials science and engineering from the National Autonomous University of Mexico, and a PhD in science (chemistry) at the University of Ghent, Belgium. Currently, he holds a research position at the Mexican Institute for Petroleum (IMP). His research areas involve organometallic, organic, and hydrothermal synthesis of materials for applications in catalysis and adsorption. He has been a visiting researcher at the University of Aberdeen, UK, working in the field of electrocatalysis. One of his main research interests is the chemical transformation of CO_2 into value-added chemicals and the production of biofuels through the application of Ionic Liquids as catalysts. He also has studied the removal of organic contaminants in fuels and the production of clean gasoline through the production of alkylate gasoline with novel catalysts. He has authored several book chapters, reviews, and research papers in the field of fuels and catalytic processes for the production of hydrocarbons and automotive gasoline. He is currently an editorial board member of the journal *Sustainable Chemistry One World*.

Contributors

Armando Balboa Palomino
National Technological Institute of
  Mexico
Technological Institute of Ciudad Madero
Tamaulipas, México

Mercedes Bazaldua Domínguez
Mexican Petroleum Institute
Mexico City, México

Silvia Castillo-Acosta
Mexican Petroleum Institute
Mexico City, México

Ricardo Cerón-Camacho
Mexican Petroleum Institute
Mexico City, México

Luis Felipe Cruz Martínez
Universidad Autónoma
  Metropolitana-Azcapotzalco
Mexico City, México

Heriberto Díaz Velázquez
Mexican Petroleum Institute
Mexico City, México

Nohra Violeta Gallardo Rivas
National Technological Institute of
  Mexico
Technological Institute of Ciudad Madero
Tamaulipas, México

Diego Guzmán-Lucero
Mexican Petroleum Institute
Mexico City, México

Javier Guzmán-Pantoja
Mexican Petroleum Institute
Mexico City, México

José Gonzalo Hernández-Cortez
Mexican Petroleum Institute
Mexico City, México

Natalya Victorovna Likhanova
Mexican Petroleum Institute
Mexico City, México

Vianey Edith López Pérez
Mexican Petroleum Institute
Mexico City, México

Rafael Martínez-Palou
Mexican Petroleum Institute
Mexico City, México

Sandra Irene Martínez Torres
National Technological Institute
  of Mexico
Technological Institute of Ciudad
  Madero
Tamaulipas, México

**María de Lourdes Mosqueira
Mondragón**
Mexican Petroleum Institute
Mexico City, México

Ulises Páramo García
National Technological Institute of
  Mexico
Technological Institute of Ciudad
  Madero
Tamaulipas, México

Araceli Vega Paz
Mexican Petroleum Institute
Mexico City, México

Preface

While central to modern civilization, the oil and chemical industry has also displayed significant environmental and sustainability challenges. However, catalytic processes offer a beacon of hope, highlighted as crucial tools for the efficient and environmentally friendly production of petroleum products, with a particular focus on automotive gasoline. This book will delve into these processes, inspiring you with their potential for a more sustainable future.

From the early steps of petroleum refining to the intricate reactions that define modern catalysts, this book offers a comprehensive journey through the principles and applications of catalysis in automotive gasoline production. It also delves into the current challenges facing this industry, such as the need for more efficient catalysts and the environmental impact of gasoline production. It explores potential solutions, including using renewable feedstocks and developing novel catalysts, to pave the way for a more sustainable future.

This book is not just a theoretical exploration but a practical guide for students and professionals in chemical engineering, materials chemistry, industrial chemistry, and related disciplines. The role in understanding and applying the catalytic processes involved in producing automotive gasoline is crucial for students in those fields, and this book is designed to equip them with the comprehensive knowledge they need to excel in their field.

Our book offers a unique perspective, covering all relevant aspects of technologies that employ catalysts to produce automotive gasoline. It not only includes classic processes in the petroleum industry but also explores unconventional and environmentally friendly processes. These include gasoline production from biomass and other non-oil sources, gasoline from greenhouse gases, mainly carbon dioxide, synthetic gasoline from methanol, and the Fischer–Tropsch process. The present and future of automotive gas and its possible survival in the face of the onslaught of electric vehicles are also discussed, providing a distinct viewpoint on the subject.

This book provides the latest information on the principles and developments in catalytic processes for gasoline production. Thus, it can serve as a good source of information for scientists and students who wish to learn more about the subject. However, the information in the book will also serve as a reasonable basis for licensees of related technologies and those involved in the oil and gas industry, including entrepreneurs who wish to start a company.

The students will learn firsthand how gasoline is currently produced since not much information in this respect is published in a document such as this book. It is also intended to be a valuable reference for researchers and scientists seeking to drive innovation in this vital field for our society's energy and environmental development.

Our book has been authored by a team of Mexican researchers from the prestigious Mexican Petroleum Institute who bring extensive experience in innovation in crude oil transformation processes. They have dedicated their expertise and knowledge to presenting you with a high-quality book. The book is structured to provide a comprehensive understanding of catalytic processes applied to gasoline production, including the historical development of catalytic processes in the petroleum industry. It explores the latest technological innovations, offering a logical and systematic journey through this fascinating field.

Welcome to a journey down the roads of catalysis toward a cleaner and more efficient future for automotive mobility!

Rafael Martínez-Palou, PhD
Researcher at the Mexican Petroleum Institute

1 Current Outlook and Future Perspective on the Production of Automotive Gasoline

Rafael Martínez-Palou
Mexican Petroleum Institute, Mexico City, México

1.1 THE CURRENT OUTLOOK OF AUTOMOTIVE GASOLINE

Currently, oil serves as the primary energy source for most countries, playing a crucial role in their economies, especially for oil-producing nations. This fossil fuel, extracted from deep within the earth, is versatile, finding use in various applications such as gasoline production, electricity generation, and the manufacturing of plastics, fertilizers, and other chemicals. However, the processes involved in oil extraction, production, transportation, and refining can have significant environmental implications.

The oil industry has been identified as a significant environmental problem due to its effects on air, water, and soil quality. The burning (combustion) of petroleum-derived fuels produces greenhouse gases (GHGs) related to global warming and atmospheric pollution. Despite the measures that have been implemented in many countries around the world to reduce the use of fossil fuels for electricity generation, the use of fossil fuels will continue to play a major role in power generation in these countries until at least the middle of this century.[1]

As the world economy recovers from the COVID-19 pandemic, demand for fuels, including gasoline, is expected to increase gradually; therefore, although the vision of fossil fuels is changing radically worldwide and there is a growing interest in migrating to new forms of environmentally friendly energy, this process will be relatively slow and at least for the next two or three decades, oil will continue to be the predominant energy source, hence the motivation for the realization of this book containing updated information on the refining processes that give rise to the hydrocarbons that will be blended in the appropriate proportion (gasoline pool).

Gasoline is in high demand, and gasoline-exporting countries obtain most of their annual profits from this item. Gasoline continues to be formulated through the formation of a pool where the hydrocarbons produced in the processes discussed in this book are mixed, such as straight-run gasoline obtained using the fluid catalytic cracking (FCC) and hydrocracking process, catalytic reforming of naphtha, isomerization/

DOI: 10.1201/9781003517283-1

"

hydroisomerization and the alkylation process, as well as the procedures to obtain additives that improve gasoline performance.[2]

Notably, the alkylation process has acquired more significant impact and contribution to the pool in the last decades not only because it is the only source of the conventional refining process that generates high-quality synthetic gasoline due to its combustion properties but also because the gasoline obtained by means of the reaction between isobutane and butenes is virtually free of the pollutants that contribute most to acid rain and environmental pollution, such as sulfur, nitrogen, and aromatic compounds.[3]

It is obviously of great interest in this book to also address new technologies for producing gasoline from renewable sources such as gasoline octane enhancers, oxygenated and nonoxygenated additives, gasoline from biomass and nonpetroleum sources, gasoline from GHGs, especially carbon dioxide (CO_2), and finally, the future of the production of fuels will be thoroughly reviewed and discussed.

Several factors, including global energy demand, crude oil prices, government policies, refining infrastructure, and the transition to cleaner, renewable energy sources, influence the current global gasoline production outlook. Gasoline demand is closely linked to economic growth and mobility.[4] Oil prices also have a significant influence, which will undoubtedly affect gasoline production and prices. Oil-producing countries, such as the Organization of the Petroleum Exporting Countries (OPEC) members and other major producers, influence prices through production and supply decisions.[5]

Crude oil refining capacity varies by region, which can affect the availability and price of gasoline in different parts of the world. Changes in refining capacity, technology upgrades, and extreme weather events can also affect gasoline production.

Government policies, such as fuel taxes, renewable energy subsidies, and emissions standards, are crucial in shaping gasoline production and consumption. In addition, increasingly stringent environmental regulations may drive the demand for more fuel-efficient vehicle technologies that are less dependent on gasoline and diesel.

The oil sector has generated significant concern about ecological problems, including effects on air and water quality, climate change, soil destruction, biodiversity losses, and its impact on the quality of human life. Moreover, oil refining is a significant source of CO_2, sulfur, and nitrogen oxides, primarily responsible for acid rain and global warming. It can also lead to soil degradation, often requiring large-scale excavation and changing the natural environment.[6]

Pipelines and associated infrastructure used to transport oil can cause soil erosion and contaminate groundwater. Oil exploration and production can damage the flora and fauna surrounding oil refineries. In addition, oil spills caused by human failures severely damage aquatic habitats and the species that live in them, affecting biodiversity.[7]

The growing awareness of climate change and the need to reduce GHG emissions, mainly by changing the current pattern of CO_2 emissions to achieve the target set in the Paris Agreement to keep global temperature rise below 2°C, has driven the transition to cleaner and renewable energy sources.[8,9]

However, it is impossible to conceive of abandoning fossil resources such as oil and gas immediately, as many industries would not be able to continue production, and there would not be enough raw materials to supply the current demand for products and services.

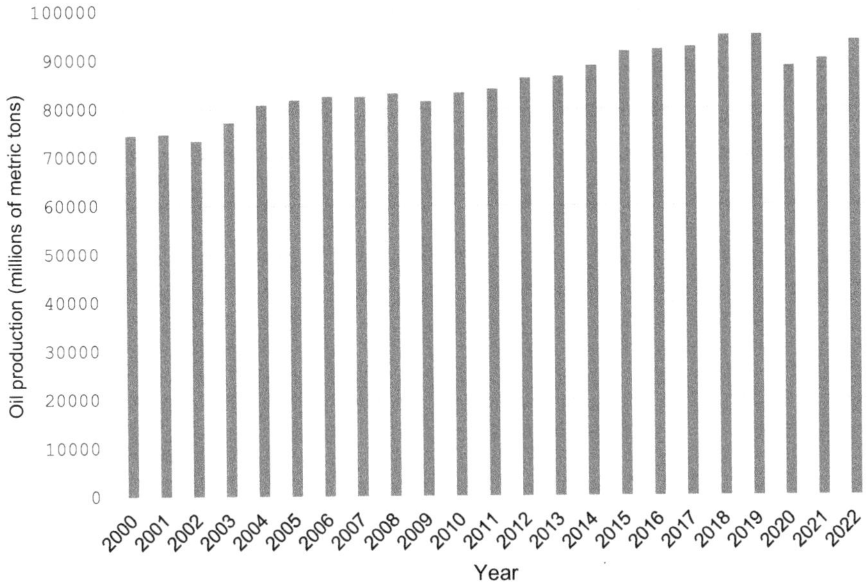

FIGURE 1.1 Global oil production 2000–2022. (Data source: Statista, Statista - the statistics portal for market data, market research and market studies).

In 2022, approximately 4.4 billion metric tons of oil were produced globally. Oil remains an important energy source, but its production must be considered in the context of sustainability and emission reductions (Figure 1.1).

As shown in Figure 1.1, the primary and relatively slight decrease occurred between 2019 and 2020 due to the COVID-19 pandemic, which caused a notable reduction in energy demand.

As a great attempt to transform the petroleum industry into a more sustainable industry, the oil industry is adopting more advanced technologies and environmentally responsible practices, such as safer drilling techniques, emergency response plans, and habitat restoration programs. Innovative methods such as 3D and 4D seismic imaging, automation and robotics, data analysis and artificial intelligence, blockchain technology, and unconventional oil extraction have revolutionized this industry, making it more efficient and with fewer adverse effects.[10]

The oil industry is continually conducting scientific and technological research to find more sustainable solutions to its environmental problems, such as the use of environmentally friendly chemical products and additives, the development of more efficient and sustainable catalysts, procedures for capturing and using toxic gases such as CO_2, sulfur and nitrogen oxides, as well as transforming them into high value-added products, and smart pipeline transition for carbon reduction as a strategic priority, including aspects such as system reforms, intelligent monitoring and control, smart performance management, and integration with the clean energy transmission grid.[11]

Table 1.1 provides an overview of the current and future trends in gasoline production worldwide, highlighting key aspects such as production, key players, environmental impact, regulations, and technology trends.

TABLE 1.1

Overview of the Current and Future Outlook for Gasoline Production

Aspect	Current Panorama	Future Trends
Global Production of gasoline	It is expected to reach approximately 100 million barrels per day (bpd) worldwide.	Moderate growth is expected but with a possible long-term decline due to the transition to cleaner energy sources.
Main Producing Countries	United States, China, Russia, India, and Saudi Arabia.	Emerging countries such as India and China are expected to increase their share of production due to rising vehicle demand and economic development.
Refining Technology	The predominant use of conventional refining technologies, with some technological innovations in processes that use highly toxic catalysts such as alkylation.	Introducing new chemical products such as additives, catalysts, and corrosion inhibitors that are more environmentally friendly (natural products, ionic liquids, etc.) is being evaluated. More efficient and ecologically friendly refining technologies, such as biomass conversion and carbon capture, are being researched and developed to reduce greenhouse gas emissions.
Environmental Impact	It contributes significantly to greenhouse gas emissions and air pollution.	A transition to electric vehicles and cleaner fuels is expected, which could reduce the demand for gasoline and its long-term environmental impacts.
Regulations and Policies	Varied regulations in different countries to limit vehicle emissions and encourage the use of cleaner technologies. Increased environmental regulations and taxes on fossil fuels are expected to promote the adoption of more sustainable alternatives.	Increased environmental regulations and taxes on fossil fuels are expected to promote the adoption of more sustainable alternatives.
International Market	Demand remains high, especially in sectors such as transportation and industry.	Demand is expected to remain stable in the short term but could gradually decline as the adoption of electric vehicles and other mobility alternatives increases.
Innovations and Technological Developments	Continued research into biofuels, synthetic fuels, and more efficient vehicle technologies. Increased investment is expected in renewable energy, energy storage, and electric mobility, which could change the energy landscape and reduce dependence on fossil fuels.	Increased investment is expected in renewable energy, energy storage, and electric mobility, which could change the energy landscape and reduce dependence on fossil fuels.[14]

Methods have been developed to obtain gasoline and diesel of fossil origin with deficient sulfur and nitrogen content, and chemical treatments required by the industry, chemical products from natural raw materials, catalysts, and nanostructured materials with low environmental impact are being developed. Applications are being developed for a family of chemical products known as ionic liquids with applications for different uses, such as catalysts, demulsifiers, corrosion inhibitors, and absorbers of sulfur, nitrogen, and aromatic compounds, among others.[12]

Researchers have also developed catalytic converters for automobiles, devices used in vehicle exhaust systems to reduce toxic gas emissions. Catalytic converters are also used in stationary engines, generators, boilers, and other industrial applications to lower GHG emissions.[13]

In the process of energy change, renewable combustion from precursors and renewable energy must be progressively increased as fossil fuels are reduced to slow global warming due to GHG emissions and reduce the risks to human health caused by increasingly severe environmental pollution.

Renewable energies are those obtained from natural resources that are inexhaustible or constantly renewable on a human scale. These energies include solar energy, which consists of harnessing solar radiation using photovoltaic or thermal solar panels to produce electricity or heat, and wind energy, which comes from the wind and is captured by wind turbines that convert the kinetic energy of wind into electricity. Hydroelectric and tidal energy is obtained by harnessing water flow, whether sea currents, rivers, or reservoirs, to move turbines that generate electricity. Geothermal energy comes from the Earth's internal heat by capturing the heat stored in the subsoil to produce electricity or heat; however, we cannot lose sight that these energies are far from meeting current energy demands.[14]

Figure 1.2 shows that in 2022, despite the great efforts made by most of the world's countries to migrate to cleaner and more sustainable energies, the percentage of energy generation from nonrenewable sources such as oil, natural gas, and coal currently stands at 81.8% compared to 14.2% for renewable energies and only 4.0% for nuclear energy, which has been losing ground as an alternative energy source due to its hazardous nature.

Renewable energies have a much lower environmental impact than fossil fuels as they do not emit GHGs or other pollutants during energy production. They also contribute to diversifying the energy matrix and reducing dependence on fossil fuels, making them essential in transitioning to a more sustainable and environmentally friendly energy system.

Biomass fuels, such as bioethanol, biodiesel, hydrotreated vegetable oil, and synthetic fuels (*synfuels*), such as ammonia or methanol, are up-and-coming alternatives to solve the environmental problem of fossil fuels; however, only if their life cycle is associated only with the use of renewable energy will they be fuels that will bring us rapidly closer to zero toxic gas emissions.[15]

In the case of biomass fuels, we cannot lose sight of the limitations regarding the use of soils other than those that can be cultivated with products intended for human consumption, the insufficiency of energy supply, particularly from renewable sources, and the lack of maturity of biofuel production technologies, which must solve

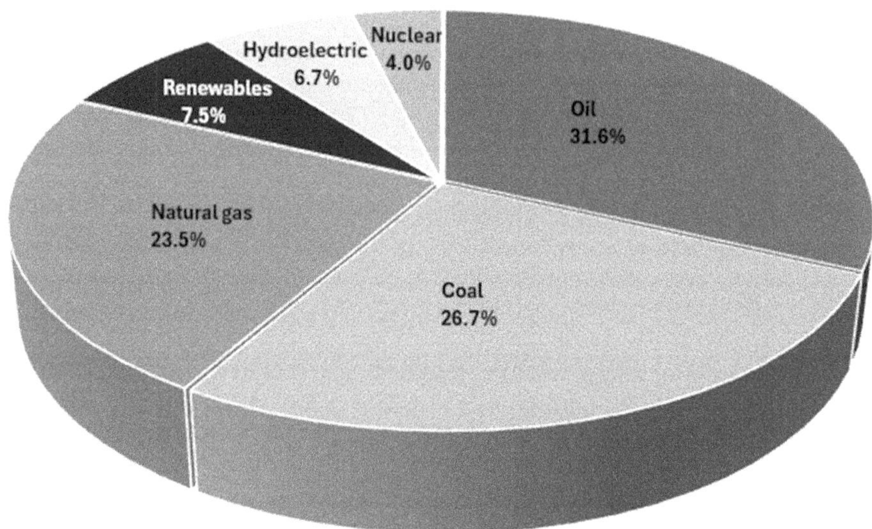

FIGURE 1.2 Global energy consumption 2022. (Data source: *Statistical review of world energy,* https://www.energyinst.org/statistical-review).

various technological problems required in the development of highly efficient biorefineries, which still require years of scientific research.

For the above reasons, although the world economy shows a slowing trend in primary energy consumption, the energy transition process will take several decades. As current statistics show, the energy consumption of fossil fuels such as gasoline, diesel, and gas has shown a sustained upward trend up to the present time, and there are no significant signs of a reversal of this trend. So, fossil fuels will still dominate the energy market for some years, if not decades.[16]

On the other hand, although renewable energies are humanity's great hope for a sustainable world, these energy sources also have their advantages and disadvantages, as shown in Table 1.2.

Currently, globally, low-carbon sources account for about 18% of the energy consumed and fossil fuels account for about 82% of the energy consumed in 2020, of which about 15% corresponds to the transportation sector.[17]

In 2024, global energy consumption is expected to grow by 1.8%, mainly due to rising demand from Asian countries. Despite high prices and unresolved supply chain disruptions, the demand for fossil fuels will reach record levels in 2024. The use of renewable energies is expected to increase by 11%.[18]

The Oil & Gas Sector Outlook 2024 explores five industry trends and drivers that are expected to play an essential role in shaping the strategies and priorities of oil and gas companies in the coming year, such as prudent capital allocation and effective execution of clean energy policies to realize the gradual energy transition and harness the growing dynamism of trade and energy relations, as well as harness the power of generative artificial intelligence for innovative solutions and new value creation.[19]

TABLE 1.2

Advantages and Disadvantages of Renewable Energies

Renewable Energy	Advantages	Disadvantages
Solar	Abundant source of free and renewable energy. Low environmental impact during operation. It can be used in remote areas without access to the power grid.	Generation depends on the availability of sunlight and can be variable depending on geographic location and weather. Requires large land areas for solar panel installation. High initial installation costs.
Wind	Renewable and abundant source of energy. Does not emit greenhouse gases or atmospheric pollutants during energy generation. Low operating costs once installed.	Wind is intermittent and not always available in all areas. Wind turbines may cause visual and noise impacts and pose risks to wildlife. Large land areas are required for wind farm installations.
Hydroelectric	Stable and predictable source of energy. Low operating and maintenance costs. Can provide large-scale energy storage through water flow regulation.	Can cause significant environmental impacts, such as altering river ecosystems and losing aquatic habitats. May require the construction of large dams, involving flooding of large land areas and displacement of communities. Vulnerable to climatic variability and drought.
Geothermal	Renewable energy source available stable and uninterruptedly. Low environmental impact during operation. Can be used for heating, cooling, and electricity generation.	Viability depends on geographic location and accessibility to geothermal resources. Geothermal drilling can pose seismic risks and release greenhouse gases. High exploration and initial development costs.
Biomass	Uses renewable organic matter as an energy source. Can utilize agricultural and forestry residues, reducing the need for landfills and deforestation. Can be used to generate electricity, heat, and biofuels.	Emits atmospheric pollutants during combustion, although to a lesser extent than fossil fuels. Biomass availability can be limited and seasonal. Large areas of arable land required Requires a constant supply of biomass to sustain energy generation.

Oil and natural gas production is expected to grow for several years, primarily to support higher energy consumption in developing Asian economies.[20]

1.2 PRODUCTION OF STRAIGHT-RUN GASOLINE

Gasoline is a mixture of combustible and volatile liquids, composed of hydrocarbons of between 5 and 12 carbon atoms and fuels derived from petroleum by atmospheric fractional distillation in a temperature range between 40 and 180°C. During fractional distillation of crude oil, the oil is separated into different fractions according to their boiling points. Straight-run gasoline is one of these fractions. Approximately

250 mL of straight-run gasoline can be obtained from each liter of crude oil during the fractional distillation process.[21]

Straight-run gasoline is one of the main components of commercial refinery, along with gasoline obtained from processes such as FCC, reforming, isomerization, and alkylation, as well as polymer gasoline, hydrocracked and coker naphtha, and thermally cracked gasoline.

Straight-run gasoline is a primary product of petroleum refining by fractional distillation of crude oil. Crude oil is heated in a furnace and vaporized in a distillation tower. As the vapor rises through the tower, it cools and condenses at different levels depending on its boiling point.

Straight-run gasoline is obtained at lower temperatures than heavier products such as diesel or lubricating oils, making it one of the lightest fractions of crude oil. However, straight-run gasoline often contains a mixture of hydrocarbons, such as alkanes, cycloalkanes, and aromatics, with different chain lengths affecting their performance and properties.[22]

Following its initial production, regular gasoline undergoes a series of additional refining processes. These steps are not just about improving its quality but they are also crucial in meeting specific requirements such as removing impurities like sulfur or adjusting its octane rating. Including processes like catalytic reforming, isomerization, and hydrotreating underscores the significance of these refining steps.

Straight-run gasoline does not have the resistance to detonation required for the proper operation of combustion engines, which is why its main application is in the chemical and petrochemical industry, where it is used to produce olefins in steam-cracking equipment.

Straight-run gasoline, in its initial state, has an octane rating of about 70, similar knocking properties as a blend of 70% isooctane and 30% heptane. However, through processes such as cracking and isomerization, the octane rating of gasoline can be increased to about 90. These processes improve the performance and efficiency of the fuel, making it more suitable for use in modern engines.

In recent decades, additives have been added to gasoline to improve its performance and antiknocking characteristics, octane index, and environmentally friendly properties, such as bioethanol, prenol, mixtures of furans (2-methylfuran and 2,5-dimethylfuran), isohexen (dimate), as well as iso-octene (di-isobutylene), methyl tert-butyl ether (MTBE), and oxygenated additives (methanol, ethanol, n-propanol, isopropanol, and di-isobutylene).[23]

The production of straight-run gasoline is a crucial step in the petroleum refining process as it provides a primary source for gasoline used in various applications, including transportation and olefins production.

1.3 ENVIRONMENTAL IMPACT OF FOSSIL GASOLINE PRODUCTION AND MITIGATION EFFORTS

It is undeniable that the production of gasoline has important environmental repercussions that range from the extraction of crude oil to the final refining processes. The extraction of crude oil entails the destruction of the habitat in surrounding areas and the loss of biodiversity, the emission of significant quantities of GHGs, contributing to climate change, spills or leaks, and environmental pollution, which can have devastating effects on marine and terrestrial ecosystems. Both the extraction and refining processes are energy-intensive, typically relying on fossil fuels, perpetuating the cycle of GHG emissions. Oil refining leads to the emission of volatile organic compounds and particulate matter, which can cause respiratory and other health problems in nearby communities.[24]

Gasoline production also has a strong impact on human health due to air and water pollution, and workers in communities near extraction sites and refineries are often exposed to pollutants that can cause respiratory problems, cancer, and other health problems.[25,26]

However, the oil industry also makes great efforts to mitigate the environmental impact of gasoline production through measures such as establishing stricter environmental regulations and continuous monitoring, which can help reduce emissions and discharges. These regulations are often implemented by policymakers who play a crucial role in overseeing and regulating the industry. Better technologies are continuously aimed at increasing efficiency and reducing harmful emissions. Applying cleaner production methods and waste management practices can minimize environmental pollution.[27]

On the other hand, great research and development efforts are made to apply more efficient and renewable chemicals and catalysts, and the transition to renewable energy sources can decrease the dependence on oil and reduce associated environmental impacts. This transition not only reduces environmental harm but also opens up new opportunities for sustainable energy production and economic growth.[28,29]

1.4 THE FUTURE PERSPECTIVE OF AUTOMOTIVE GASOLINE

The future of gasoline worldwide is subject to several factors and trends shaping the global energy landscape. Many countries are promoting policies for the mass adoption of electric vehicles to reduce GHG emissions and dependence on fossil fuels.[30]

To reduce emissions to zero, heavy-duty vehicles and public and private transport must migrate virtually 100% to electric vehicles, for which the cost of electric vehicles must be significantly reduced. The subsidy for these vehicles must be implemented or increased, which is expected to happen as battery technology improves.[31]

In addition to electric vehicles, alternative fuel technologies, such as hydrogen, bioethanol, and others that could compete with gasoline in the future, are also being developed. They are already being used on a small scale, and bioethanol, in particular, is becoming more widely used in low-percentage blends with gasoline.[32]

Many countries are implementing policies to reduce carbon emissions (decarbonization policies), which may include carbon taxes, stricter regulations on vehicle emissions, and subsidies for clean technologies. These policies could affect gasoline demand and promote more sustainable alternatives. If additional measures arc not implemented and compliance mechanisms are not made more aggregated, the percentages described above would not be sufficient to ensure the timely decarbonization of the light-duty vehicle fleet.[33]

Another strategy that is gradually being implemented, prompted by increasing urbanization and concerns about traffic congestion and pollution, is changes in urban mobility, such as encouraging public transport and carpooling and promoting more sustainable modes of transport, such as cycling and active transportation. These changes can reduce gasoline demand in urban environments.[34]

While gasoline demand is likely to decline in the future due to factors such as the adoption of electric vehicles and decarbonization policies, it is unlikely to disappear entirely in the short term. However, the oil and gas industry is likely to adapt and diversify its operations to cope with these changes in the energy market.[35]

Gasoline production is closely tied to crude oil production, as gasoline is refined from crude oil. Changes in global oil production will influence the availability and cost of gas. Advances in refining technology and increased regulation of emissions from fossil fuels, including gasoline, may also improve efficiency and reduce gasoline production costs. This could affect both gasoline production and consumption and make gasoline production more profitable and sustainable in the future.

While gasoline production remains an integral part of the global economy today, its future is subject to several forces and trends; this could result in a gradual decline in gasoline production and consumption over the long term, although demand for fossil fuels is likely to persist to some extent for some time.

Although the costs of using sustainable fuels may be higher in the long term, the using 100% renewable diesel and hydrotreated vegetable oil (biofuel jet) could achieve GHG reductions comparable to those of electric vehicles, enabling more rapid decarbonization of existing fleets in the short term.[36]

1.5 CONCLUSIONS

As we have seen throughout the introductory chapter, fossil fuels will continue to play a major role in power generation for at least three decades to come. Although the world is undergoing a process of energy transition to cleaner renewable fuels, this process will be gradual. The future of gasoline in the world will be determined by a number of factors and trends shaping the global energy landscape, and the energy transition process will depend on each individual country and its commitment to emission reduction.

REFERENCES

(1) Abas, N.; Kalair, A.; Khan, N. Review of Fossil Fuels and Future Energy Technologies. *Futures.* 2015, *69*, 31–49. https://doi.org/10.1016/j.futures.2015.03.003

(2) Velázquez, H. D.; Cerón-Camacho, R.; Mosqueira-Mondragón, M. L.; Hernández-Cortez, J. G.; Montoya de la Fuente, J. A.; Hernández-Pichardo, M. L.; Beltrán-Oviedo, T. A.; Martínez-Palou, R. Recent Progress on Catalyst Technologies for High Quality Gasoline Production. *Catal. Rev. - Sci. Eng.* 2023, *65* (4), 1079–1299. https://doi.org/10.1080/01614940.2021.2003084

(3) Guzmán-Lucero, D.; Guzmán-Pantoja, J.; Velázquez, H. D.; Likhanova, N. V.; Bazaldua-Domínguez, M.; Vega-Paz, A.; Martínez-Palou, R. Isobutane/Butene Alkylation Reaction Using Ionic Liquids as Catalysts. Toward a Sustainable Industry. *Mol. Catal.* 2021, *515*, 111892. https://doi.org/10.1016/j.mcat.2021.111892

(4) *Energy outlook* 2024. Economist Intelligence Unit. https://www.eiu.com/n/campaigns/energy-in-2024/ (accessed 2024-03-12).

(5) https://Www.Opec.Org/Opec_web/En/.OPECEnergyRev

(6) Katopodis, T.; Sfetsos, A. A Review of Climate Change Impacts to Oil Sector Critical Services and Suggested Recommendations for Industry Uptake. *Infrastructures* 2019, *4* (4), 74. https://doi.org/10.3390/infrastructures4040074

(7) Ritchie, H.; Rosado, P.; Roser, M. *CO_2 and Greenhouse Gas Emissions*. Our World Data 2023.

(8) Muradov, N. Low to Near-Zero CO_2 Production of Hydrogen from Fossil Fuels: Status and Perspectives. *Int. J. Hydrog. Energy.* 2017, *42* (20), 14058–14088. https://doi.org/10.1016/j.ijhydene.2017.04.101

(9) Hoang, A. T.; Foley, A. M.; Nižetić, S.; Huang, Z.; Ong, H. C.; Ölçer, A. I.; Pham, V. V.; Nguyen, X. P. Energy-Related Approach for Reduction of CO_2 Emissions: A Critical Strategy on the Port-to-Ship Pathway. *J. Clean. Prod.* 2022, *355*, 131772. https://doi.org/10.1016/j.jclepro.2022.131772

(10) Alagoz, E.; Alghawi, Y.; Ergul, M. S. Innovation in Exploration and Production: How Technology Is Changing the Oil and Gas Landscape. *J. Energy Nat. Resour.* 2023, *12* (3), 25–29. https://doi.org/10.11648/j.jenr.20231203.11

(11) Liao, Q.; Liang, Y.; Tu, R.; Huang, L.; Zheng, J.; Wang, G.; Zhang, H. Innovations of Carbon-Neutral Petroleum Pipeline: A Review. *Energy Rep.* 2022, *8*, 13114–13128. https://doi.org/10.1016/j.egyr.2022.09.187

(12) Martínez-Palou, R.; Likhanova, N. V. *Applications of Ionic Liquids in the Oil Industry: Towards A Sustainable Industry*; Bentham Science Publishers, 2022. https://doi.org/10.2174/97898150795791230101

(13) Robles-Lorite, L.; Dorado-Vicente, R.; Torres-Jiménez, E.; Bombek, G.; Lešnik, L. Recent Advances in the Development of Automotive Catalytic Converters: A Systematic Review. *Energies.* 2023, *16* (18), 6425. https://doi.org/10.3390/en16186425

(14) Mohtasham, J. Review Article-Renewable Energies. *Energy Procedia.* 2015, *74*, 1289–1297. https://doi.org/10.1016/j.egypro.2015.07.774

(15) International Energy Agency. *Net Zero by 2050 – Analysis*. IEA. https://www.iea.org/reports/net-zero-by-2050 (accessed August 05, 2024).

(16) Energy Institute. *Statistical review of world energy*. Statistical review of world energy. https://www.energyinst.org/statistical-review/home (accessed July 03, 2024).

(17) *Renewables – Global Energy Review 2021 – Analysis*. IEA. https://www.iea.org/reports/global-energy-review-2021/renewables (accessed August 05, 2024).

(18) *Energy outlook 2024*. Economist Intelligence Unit. https://www.eiu.com/n/campaigns/energy-in-2024/ (accessed August 05, 2024).

(19) *2024 oil and gas industry outlook*. Deloitte Insights. https://www2.deloitte.com/us/en/insights/industry/oil-and-gas/oil-and-gas-industry-outlook.html (accessed August 05, 2024).

(20) Chen, H.; He, J.; Zhong, X. Engine Combustion and Emission Fuelled with Natural Gas: A Review. *J. Energy Inst.* 2019, *92* (4), 1123–1136. https://doi.org/10.1016/j.joei.2018.06.005

(21) Hsu, C. S.; Robinson, P. R. Gasoline Production and Blending. In *Springer Handbook of Petroleum Technology*; Hsu, C. S., Robinson, P. R., Eds.; Springer International Publishing: Cham, 2017; pp 551–587. https://doi.org/10.1007/978-3-319-49347-3_17

(22) Erofeev, V. I.; Khomyakov, I. S.; Egorova, L. A. Production of High-Octane Gasoline from Straight-Run Gasoline on ZSM-5 Modified Zeolites. *Theor. Found. Chem. Eng.* 2014, *48* (1), 71–76. https://doi.org/10.1134/S0040579514010023

(23) Abdellatief, T. M. M.; Ershov, M. A.; Kapustin, V. M.; Ali Abdelkareem, M.; Kamil, M.; Olabi, A. G. Recent Trends for Introducing Promising Fuel Components to Enhance the Anti-Knock Quality of Gasoline: A Systematic Review. *Fuel* 2021, *291*, 120112. https://doi.org/10.1016/j.fuel.2020.120112

(24) Murphy, R. What Is Undermining Climate Change Mitigation? How Fossil-Fuelled Practices Challenge Low-Carbon Transitions. *Energy Res. Soc. Sci.* 2024, *108*, 103390. https://doi.org/10.1016/j.erss.2023.103390

(25) Perera, F.; Ashrafi, A.; Kinney, P.; Mills, D. Towards a Fuller Assessment of Benefits to Children's Health of Reducing Air Pollution and Mitigating Climate Change Due to Fossil Fuel Combustion. *Environ. Res.* 2019, *172*, 55–72. https://doi.org/10.1016/j.envres.2018.12.016

(26) Perera, F. Pollution from Fossil-Fuel Combustion Is the Leading Environmental Threat to Global Pediatric Health and Equity: Solutions Exist. *Int. J. Environ. Res. Public. Health.* 2018, *15* (1), 16. https://doi.org/10.3390/ijerph15010016

(27) Arthur, J. L. Mitigating the Environmental Effects of Oil and Gas Exploitation: Issues of Compliance, Cost of Production, and Community Awareness. *J. Power Energy Eng.* 2020, *8* (9), 51–64. https://doi.org/10.4236/jpee.2020.89005

(28) Hayashi, S. H. D.; Ligero, E. L.; Schiozer, D. J. Risk Mitigation in Petroleum Field Development by Modular Implantation. *J. Pet. Sci. Eng.* 2010, *75* (1), 105–113. https://doi.org/10.1016/j.petrol.2010.10.013

(29) Swick, D.; Jaques, A.; Walker, J. C.; Estreicher, H. Gasoline Risk Management: A Compendium of Regulations, Standards, and Industry Practices. *Regul. Toxicol. Pharmacol.* 2014, *70* (2, Supplement), S80–S92. https://doi.org/10.1016/j.yrtph.2014.06.022

(30) Muradov, N. Low to Near-Zero CO_2 Production of Hydrogen from Fossil Fuels: Status and Perspectives. *Int. J. Hydrog. Energy* 2017, *42* (20), 14058–14088. https://doi.org/10.1016/j.ijhydene.2017.04.101

(31) Kalghatgi, G. Is It Really the End of Internal Combustion Engines and Petroleum in Transport? *Appl. Energy.* 2018, *225*, 965–974. https://doi.org/10.1016/j.apenergy.2018.05.076

(32) Teixeira, A. C. R.; Machado, P. G.; de Collaço, F. M. A.; Mouette, D. Alternative Fuel Technologies Emissions for Road Heavy-Duty Trucks: A Review. *Environ. Sci. Pollut. Res.* 2021, *28* (17), 20954–20969. https://doi.org/10.1007/s11356-021-13219-8

(33) Olkkonen, V.; Hirvonen, J.; Heljo, J.; Syri, S. Effectiveness of Building Stock Sustainability Measures in a Low-Carbon Energy System: A Scenario Analysis for Finland until 2050. *Energy.* 2021, *235*, 121399. https://doi.org/10.1016/j.energy.2021.121399

(34) Zhu, Z.; Chen, B.; Chen, H.; Qiu, S.; Fan, C.; Zhao, Y.; Guo, R.; Ai, C.; Liu, Z.; Zhao, Z.; Fang, L.; Lu, X. Strategy Evaluation and Optimization with an Artificial Society toward a Pareto Optimum. *The Innovation.* 2022, *3* (5). https://doi.org/10.1016/j.xinn.2022.100274

(35) Garcia, R.; Freire, F. A Review of Fleet-Based Life-Cycle Approaches Focusing on Energy and Environmental Impacts of Vehicles. *Renew. Sustain. Energy Rev.* 2017, *79*, 935–945. https://doi.org/10.1016/j.rser.2017.05.145

(36) Harvey, L. D. D. Rethinking Electric Vehicle Subsidies, Rediscovering Energy Efficiency. *Energy Policy* 2020, *146*, 111760. https://doi.org/10.1016/j.enpol.2020.111760

2 Catalytic Cracking

María de Lourdes Mosqueira Mondragón
and Rafael Martínez-Palou
Mexican Petroleum Institute, México City, México

2.1 INTRODUCTION

The transformation of petroleum into products useful to humans is not new; its origin dates back to the intersection of politics, society, and technology at the end of the 19th century. Although there are records of petroleum-derived materials being used centuries ago in various parts of the world, either as fuel to light lamps or as street filler, it was not until the invention of internal combustion engines that the increasing importance of petroleum and its uses was realized. This increase was mainly evidenced during the Second World War, where its use as fuel (gasoline, diesel, turbosine, among others) was crucial.

In addition to its function as a fuel, oil is used in the production of proteins (arginine, lysine, leucine, isoleucine, valine, methionine, threonine, among others), low-boiling liquid products, plastics, synthetic fibers, such as nylon, acrylic, and polyester, fine chemical solvents, pharmaceuticals, waxes, and coke. This is due to the versatile properties of oil, which make it an indispensable resource for a population that is growing day by day and requires specific needs to be met.[1–4]

Currently, the composition of oil is more complex, presenting compounds of several thousand atomic mass units, as well as higher contents of heteroatoms such as Ni, V, O, S, and N, which must be removed. This implies the optimization of existing fuel refineries due to the continuous increase in demand, without affecting the environment. Today, refineries face the challenge of obtaining high-quality fuels from depleted light crude sources and consider alternatives such as the use of heavy oils, cracked gas oils, deasphalted gas oils, and atmospheric vacuum residues. The purpose of these processes is to reduce viscosity and boiling point, as well as to carry out demetallization, desulfurization, denitrogenation, and increase the H/C ratio, providing added value to the material obtained.[5,6]

To achieve this, several catalyst synthesis processes and heavy crude processing technologies have been developed, including hydrocracking, which considers feeds with high impurities. Metal catalysts perform disintegration and hydrogenation to

DOI: 10.1201/9781003517283-2

obtain low boiling point distillates or middle distillates with high-value products. The hydrogenation exothermic reaction is carried out on olefin compounds, aromatics, sulfur, nitrogen, and oxygen, while the disintegration reaction, endothermic, is carried out on the C–C bonds, generating net heat in the reactor.

Among the valuable products obtained in this process is gasoline, a mixture of flammable hydrocarbons obtained by fractional distillation of crude oil and used in internal combustion engines. Its use in the daily life of today's inhabitants of the planet is due to its volatility, flammability, and octane rating properties, which provide advantages such as easy cold starting and maximum power during vehicle acceleration.[7–9]

There are several processes to produce gasoline and light distillates, such as thermal disintegration at high temperatures (850 °C or higher) and high pressures (620 kPa), and the Shukhov cracking process in batch reactors, which has been carried out since 1891.[10,11]

The Houdry catalytic disintegration is carried out in a fixed bed using several acid clays (crystalline aluminosilicates) to disintegrate large molecules, generating large quantities of coke that had to be removed in a subsequent process to allow regeneration and recirculation of the catalysts to the reactor. The raw material was the same as that used in a thermal disintegrator, that is, light diesel oil, from which gasoline was produced. Subsequently, fluidized beds and the use of crystalline silico-aluminates emerged, and environmental restrictions forced the design of environmentally friendly processes, proliferating to become a fundamental component of the modern oil refining process.[5,12]

Thus arose the fluid catalytic cracking (FCC) process, which is the catalytic disintegration of oil (catalytic cracking) and is carried out in a fluidized bed reactor, being one of the most important conversion processes in a refinery that marked a substantial advance in the middle of the 20th century. At that time, the Standard Oil Company® of New Jersey successfully commissioned the first fluidized catalytic cracking (FCC) unit. Since then, the use of the FCC process has grown exponentially in the field of catalytic cracking. It is an innovative technology in the petrochemical and refining industry. There are more than 300 FCC units worldwide processing more than 14.4 million barrels per day (BPD) and utilizing the latest technology in this important field. To date, there are more than 8,200 papers published during the last 10 years alone, and about 45% of all gasoline worldwide comes from FCC units.[13]

The high yield and selectivity of the catalytic disintegration reaction depend on several factors such as the type of feed, the operating conditions of the process (temperature, residence time, and catalyst/oil ratio, for example), and an effective catalyst. For the FCC process, the catalysts employed are generally synthesized with a zeolite, a matrix, and a binder.

The FCC process is highly complex but also versatile and cost-effective. The unit consists of a reactor section and a regenerator section connected by transfer lines that allow free transport of spent and regenerated cracking catalysts between them.[13] A schematic of the FCC process is shown in Figure 2.1.

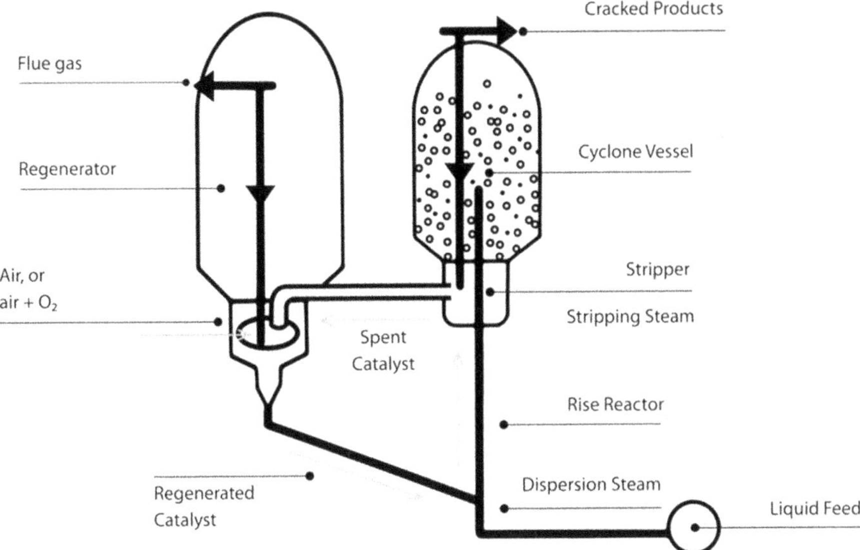

FIGURE 2.1 Diagram of an FCC process unit.

2.2 CATALYST FOR THE FCC PROCESS

Catalyst design is critical in the FCC process. Catalytic disintegration is a complex process that requires experimental information and mathematical models, some very complicated, to express the input and output variables in industrial practice.

FCC catalysts have evolved over the years since their initial commercial application. The history of disintegration catalysts began with the use of natural clays, specifically bentonites, with relatively high acidity, low activity, and low selectivity. Despite having properties such as adequate specific surface area and pore volume, layered structure, cation exchange capacity, and being environmentally friendly, non-toxic, and low-cost raw materials, these clays were rapidly deactivated, which led to the investigation of other materials such as silica and/or alumina to improve the catalytic activity.[4] However, these synthesized materials proved to be unstable and difficult to regenerate despite their good activity.

The development of a catalyst for the FCC process with high resistance to contaminants has been a challenge in its optimization. The transformation of heavy molecules with high pollutant contents requires feed pretreatments. In addition, diesel and gasoline as valuable products need catalysts with mesoporous and microporous structures, suitable acidity, and hydrothermal stability, among other properties. This must consider elevated temperatures due to extreme conditions and attrition wear in the reactor and the regenerator, among other factors.

In 1962, the Y-type zeolite-based catalyst (FAU) became a commercial catalyst. Developed by Mobil Corporation and Union Carbide, this catalyst had high activity and selectivity for light petroleum fractions.[11]

The zeolites of the FCC process have micropores that are usually smaller than 1 nm, which limits the diffusion of compounds within these pores and the size of the molecules that can be catalyzed. To improve the internal accessibility of the pores, techniques such as desilication with alkaline solutions, high-temperature vaporization, or chemical treatments have been developed. However, these techniques can alter the structure and composition of zeolites, affecting the acidity, Si/Al ratio, and uniformity of mesoporosity.

Other procedures using templates have also been developed at a higher cost.[14] In addition, the hydrothermal stability of large-pore zeolites is improved by controlled hydrothermal treatments combined with acid washing cycles, the incorporation of alkali and selected transition metals and rare earth metal ion exchange or a combination of both, or techniques such as P incorporation.[15,16] In the last 50 years, the FCC catalyst contains ultra-stabilized zeolite Y (USY) deficient in Al[1], although structures such as MCM-41 with a mesoporous structure have also been considered, but they have a lower mechanical stability.

In the last 50 years, the FCC catalyst has evolved into the Al-deficient USY, although structures such as MCM-41 with a mesoporous structure, which have lower mechanical stability, have also been considered.

Zeolitic materials have undergone significant evolution to maximize the conversion of complicated feedstocks into valuable transportation fuels and light olefins with improved technology that retains their activity despite contaminants and with high conversions. Technologies with a mixture of catalysts performing independent functions or co-catalysts have been proposed, all aimed at addressing the problems associated with heavier feedstock complexity and higher contaminant content.[9,11,13,17,18]

Several modified forms of zeolite Y have been developed, for example, rare earth exchanged Y (REY) such as La and Ce, which have been employed in vacuum gas oil (VGO) cracking, and its saturated (SF), aromatic (AF), and resinous (RF) fractions, stabilizing the zeolite structure and increasing its thermal and hydrothermal stability.[19,20]

Subsequently, another important catalyst that appeared is the ZSM-5 zeolite, which also disintegrates and is mainly applied as an additive to the conventional FCC catalyst to improve the selectivity of light olefins, such as propylene, and increase the octane number of gasoline. Zeolites have been steadily improved to date, including modification of particle size and acidity, phosphorus treatment, and creation of hierarchical mesoporous structures, among other treatments.

Recently, new zeolites have been tested in the FCC process instead of traditional zeolite Y as disintegration catalysts. Catalysts for the FCC process are generally synthesized with zeolite, a matrix, a filler, a binder, and certain metals for specific functions. Thus, a typical FCC process catalyst may contain an active ingredient (10 to 50% by weight) dispersed in a solid matrix (50 to 90% of the total), which provides physical and mechanical resistance and incorporates the active component and additives that confer tolerance to the catalyst to deactivation by poisoning, always seeking economy and process optimization to transform increasingly heavy fossil wastes into valuable products such as gasoline. Figure 2.2 shows the main components of the catalyst formulations for the FCC process.

The zeolite structure consists of a network of very small pores (~8.0 Å in diameter), with an average surface area of 600 m^2/g[21]. This means that hydrocarbon

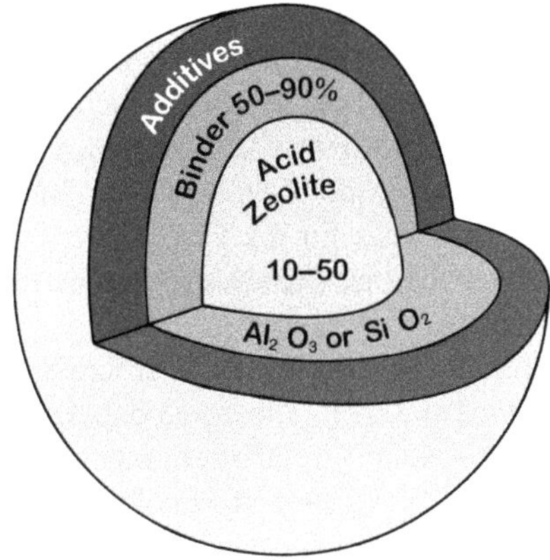

FIGURE 2.2　Main components of the catalyst formulations for the FCC process.

TABLE 2.1
Zeolites Are Used in the FCC Process

Function in the Process	Zeolite	Products
Catalyst	HY[22]	Gasoline (branched hydrocarbons), kerosene, and gas
Catalyst	USY[23], Mesoporous USY[22], REY[24], HZSM-5[25], ZSMH-5/Y[26], P-HZSM-5[27], ZSM-5/SBA-15[28], H-Ferrierite[29], Mordenite[29], IM-5 Mesoporous[30], and Natural zeolite[31]	Gasoline
Catalyst	HZSM-5[22]	Light olefins and gasoline
Catalyst	H-Beta[22]	Gasoline
Catalyst	Me/Y[29]	Middle distillate
Catalyst	Me/ZSM-5[32]	Biogasoline
Additive	H-ZSM-5[25], SVR[33], ITH[33], and MFI[33]	Gasoline

RE: Rare earth (La) Me: Metals like Fe, Zn, Cu, and Zr.

molecules with diameters larger than 8.0 Å cannot be admitted into the pores, which prevents mass transfer. Some zeolites commonly used in FCC process reactions are presented in Table 2.1.

These materials are synthesized as fine powders with a particle size of 70–75 μm and a crystal size <1 μm. Particles outside this range are discarded to avoid being retained by the FCC unit if they are too small or may cause problems in fluidization if they are too large. In addition, they must possess high activity and selectivity and achieve high reagent diffusivity to convert large molecules into molecules of lower molecular weight, as well as resistance to friction due to fluidized bed operation, hydrothermal stability, resistance to reaction temperature and vapor pressure during regeneration, corrosion resistance, reaction temperature, tolerance to heavy feed metals (which deposit on the active sites, poisoning the catalyst), and low selectivity to coke (produced by the high cracking activity, which increases with heavy crude oil), properties that are achieved with the use of zeolites.

Globalization and growth in consumption levels have resulted in the difficult and limited availability of raw materials for the FCC process. The performance of the FCC unit products, as well as their properties, depends on the feedstock and is especially critical when processing heavy fuels.[34,35]

The feedstock for the FCC process contains different amounts of heteroatoms depending on the origin of the crude oil. The most common contaminating metals are vanadium, nickel, iron, and sodium.[1] In addition, other metals such as Zn, Pb, Cu, Cd, Cr, Co, As, Sb, Te, Hg, Au, or Ag, in the form of metallic porphyrins, are present in crude oils worldwide, mainly in Venezuela and Mexico.[36,37]

Changes in the quality of petroleum feedstocks significantly affect catalyst deactivation in catalytic cracking, resulting in the formation of coke on the catalyst surface. Factors such as saturates, aromatics and resins (SAR) content, API gravity, molecular weight, aniline point, refractive index, bromide index, density, and viscosity in the feedstocks, as well as major process variables such as temperature and pressure, influence coke formation on the catalyst and loss of catalytic activity.[38,39]

Optimization of advanced catalysts and pretreatment of FCC process feedstocks drive different technologies applied to process VGO, VGO mixed with residual oil, coke oil, deasphalted oil, and atmospheric residue.[40,41]

Various strategies have been tried for processing heavy fuels, such as atomization of the feedstock, which allows instantaneous vaporization and rapid cooling of the catalyst.[42] Other methodologies include treatments prior to the FCC process, such as solvent deasphalting or deep solvent prefractionation by extraction and supercritical fluids.[43,44]

Knowing the composition and properties of the feed has a direct impact on the octane rating of the gasoline blend product, which is a function of the concentration of the various hydrocarbons in the blend. Aromatic compounds have the highest-octane rating, while n-paraffins have the lowest.[5] However, the current trend is to reduce the content of aromatic compounds due to the health problems they can cause.[45]

Pretreatment of feedstocks for the FCC process is vital to reduce the content of impurities such as S, N, and metals in the feed to the cracking unit, as well as to decrease the viscosity and molecular weight through some pretreatment hydrotreatment, which contributes to reducing pollutant emissions to the atmosphere.

A detailed knowledge of the structure and composition of the feedstock for the FCC process allows for identifying, contributing, and solving problems in the catalyst and the operating conditions through an analysis of the reaction routes for the generation of the products of interest.

2.3 REACTION OF THE FCC PROCESS

FCC has the great advantage of producing large quantities of light distillate without adding hydrogen. In the FCC process, a complex network of chemical reactions converts high-boiling hydrocarbons into valuable gasoline and other lighter products.

When heavy hydrocarbons are passed through a catalyst at high temperatures and moderate pressures, they undergo cracking that breaks the larger molecules into smaller molecules.

Heavy hydrocarbons ⟶ Lighter hydrocarbons + Catalyst regeneration

This process involves breaking the carbon–carbon bonds of the larger molecules, resulting in a mixture of smaller hydrocarbons, including alkenes and aromatic compounds, which are separated and further processed to obtain the desired products. The catalyst used in the process facilitates this cracking reaction and is continuously regenerated to maintain its activity.

Some of the most important reactions are:

Cracking: This is the main reaction in the FCC process, in which large hydrocarbon molecules are broken down into smaller, more valuable products. The cracking reaction usually occurs via a free radical mechanism, which involves breaking carbon–carbon bonds within the hydrocarbon molecules. This reaction produces a mixture of smaller hydrocarbons, such as olefins, kerosenes, and aromatic compounds. An example of this reaction is given in Equation 2.1.

$$\text{(2.1)}$$

Isomerization: During the cracking process, some of the hydrocarbons produced may have branched or cyclic structures. Isomerization reactions convert these molecules into straight-chain or more stable isomers, increasing the final product's quality by improving its octane rating. An example of this reaction is given in Equation 2.2.

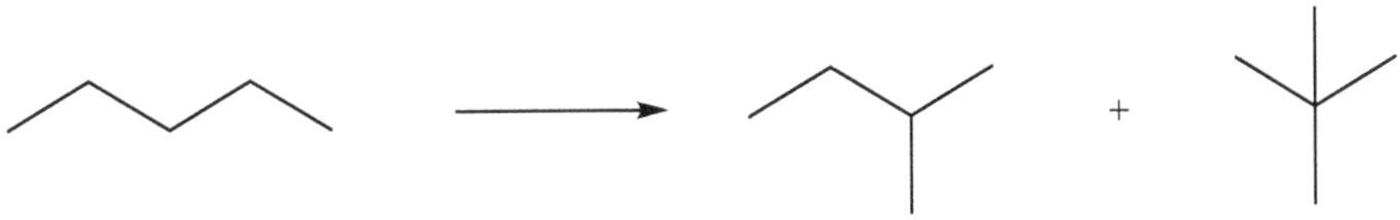

$$\text{(2.2)}$$

Hydrogen transfer: Hydrogen transfer reactions occur simultaneously with cracking and isomerization reactions. These reactions involve the transfer of hydrogen atoms between hydrocarbon molecules, leading to the formation of more stable and saturated compounds.

An example reaction is as follows:

$$+ \ H_{2\,(g)} \longrightarrow \qquad\qquad \text{(2.3)}$$

Coke formation: In addition to producing the desired products, cracking reactions can also lead to the formation of coke, which is a carbonaceous residue. Coke formation occurs when the cracking process is incomplete or when conditions favor carbon deposition on the catalyst surface. Coke deposition deactivates the catalyst and reduces its catalytic efficiency, making it necessary to regenerate it.

These reactions occur in the presence of a catalyst, usually a zeolite-based material with acidic properties. The catalyst facilitates the cracking process by providing active sites for the adsorption and transformation of hydrocarbon molecules. The FCC process operates at elevated temperatures (~490–570°C) and moderate pressures, typically in a fluidized bed reactor system, to ensure effective mixing and contact between the hydrocarbon feedstock and catalyst particles.

2.4 CRACKING REACTIONS MECHANISM

2.4.1 THERMAL CRACKING

Cracking reactions can occur by thermal or catalytic cracking. When the temperature in the system is higher than 800°C, hydrocarbon thermal cracking (pyrolysis) can occur by a free-radical mechanism.[46] This mechanism involves the homolytic cleavage of carbon–carbon bonds in hydrocarbon molecules, forming free radicals that react with other molecules to form smaller cracking products. A homolytic carbon–carbon break occurs in this process, forming products with broad molecular weight distribution different from those obtained by conventional catalytic cracking.[47–49]

The homolytic cracking mechanism promotes the formation of ethylene. The following reactions can simplify the steam-cracking reaction of hydrocarbons by free radicals (Figure 2.3).

Termination reactions can occur when two free radicals react with each other to generate a stable product or when one free radical transfers a hydrogen atom to another to produce a saturated and unsaturated hydrocarbon (Figure 2.4).

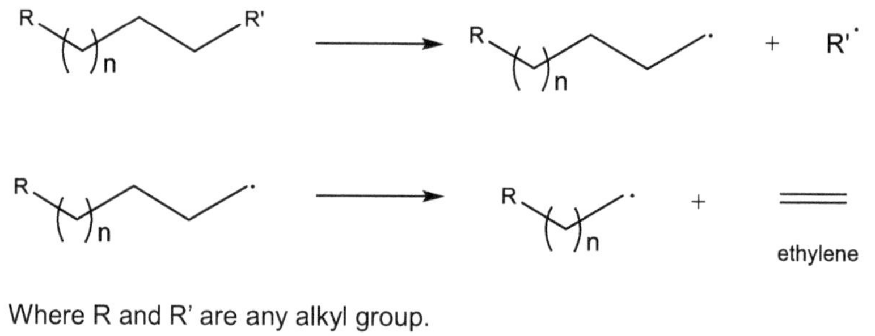

FIGURE 2.3 Free-radical mechanism (unimolecular).

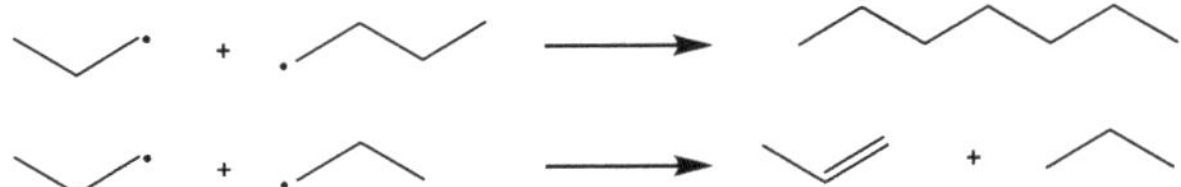

FIGURE 2.4 Termination reactions in the presence of free radicals.

2.5 CATALYTIC CRACKING

In general, cracking reactions occur in three stages: Initiation, propagation, and termination. The catalytic cracking predominantly occurs through the formation of carbocation mechanisms by heterolytic breaking of C–C bonds. In the carbocation formation mechanism, conversely, the heterolytic cleavage of carbon–carbon bonds results in the formation of positive ions or carbocations.[50] These carbocations are more stable intermediates than free radicals that can rearrange proton transfers or alkyl group displacements to form cracking products.

It is widely accepted that the catalytic cracking mechanism classically takes place through a bimolecular reaction (hydride (H-) transfer) and tends to produce more selective products with narrower molecular weight distributions than the free radical mechanism. According to this mechanism, a carbenium ion (hydrocarbon ions having a positive charge on a carbon atom) is formed by hydride transfer or β-scission (Figure 2.5).[51]

As seen in Figure 2.5, this mechanism occurs when the β-scission of secondary and/or tertiary alkylcarbenium ions occurs to give rise to smaller alkylcarbenium ions and alkenes. The abstraction of a hydride ion from a tertiary carbon is easier than from a secondary carbon, and even easier than from a primary carbon. Subsequently, bimolecular hydride transfer reactions involving alkylcarbenium ions and alkenes produce high yields of short-chain alkanes and lower yields of olefins.

Another mechanism of heterolytic cleavage of C–C bonds accepted the mechanism of protolytic cracking of alkanes proposed by Haag and Dessau.[52] This mechanism is favored for medium pore zeolite catalysts that facilitate the monomolecular reaction of alkanes.

In this mechanism, an acid catalyst protonates an alkane to give the transition states of carbonium ions that collapse to give alkanes and carbene ions, which return the protons to the catalyst to form alkenes. This cracking promotes the formation of small molecules such as hydrogen, methane, and ethane (Figure 2.6).[52]

The mechanism for the process using zeolites as catalysts has been studied by several research groups,[47,53–55] and it is recognized that the reaction mechanism runs according to the following stages:

$$RH + R'^{+} \xrightarrow{\text{H· transfer}} R^{+} + R'H$$

$$R^{+} \xrightarrow{\text{β-scission}} R'^{+} + \text{olefin}$$

FIGURE 2.5 Mechanism of bimolecular hydride transfer reactions involving alkylcarbenium.

$$RH + H^+ \xrightarrow{\text{carbonium ion formation}} RH_2^+$$

$$RH_2^+ \xrightarrow{\text{cracking}} R'H + R''H^+$$

FIGURE 2.6 Protolytic mechanism of catalytic cracking

1. The zeolite through its Lewis acid sites abstracts a hydride ion from a hydrocarbon.

$$(2.4)$$

2. Rearrangement of carbenium ion by hydride ion shift to generate a more stable cation (isomerization).

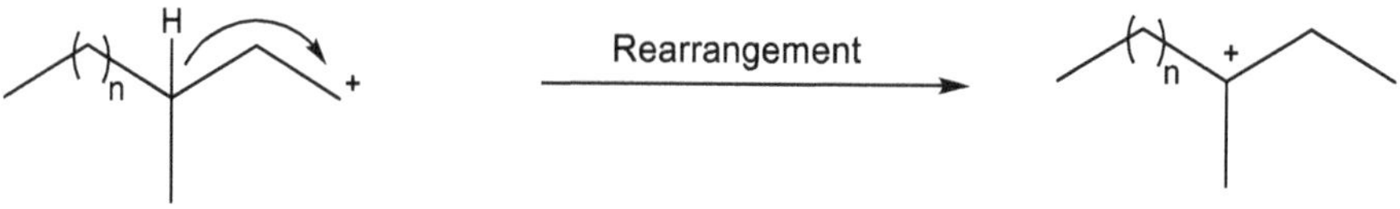

$$(2.5)$$

3. Reaction of an olefin with a zeolite proton ion (Brønsted site) to generate a carbenium ion (Eq. 2.6).

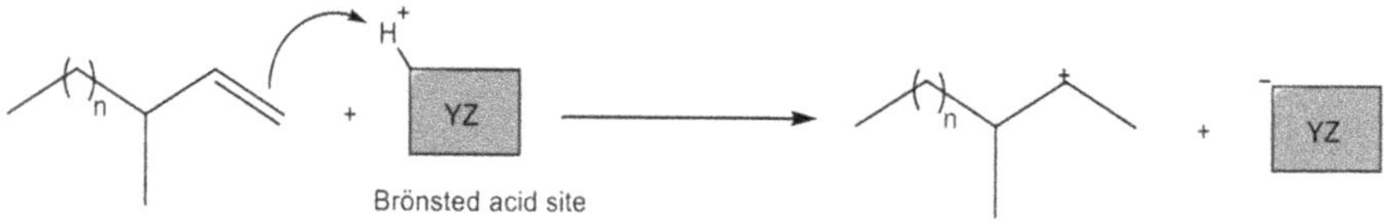

$$(2.6)$$

4. Rearrangement by 1,2-methyl shift (isomerization, Eq. 2.7).

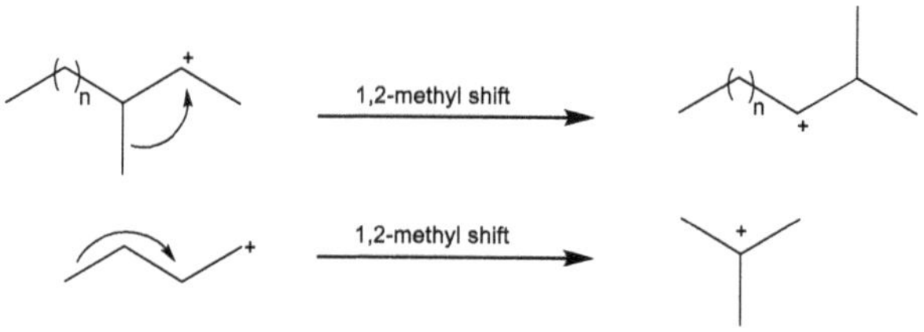

$$(2.7)$$

5. Reaction of a carbenium ion with a hydrocarbon by abstraction of a hydride ion and cracking to generate lighter carbenium ion (Eq. 2.8).

$$(2.8)$$

6. Reaction propagation by β-scission of the new carbenium ions to yield olefine and lighter hydrocarbons (Eq. 2.9).

$$(2.9)$$

7. Termination reactions

Finally, termination reactions can occur when a carbenium ion abstracts a hydride ion from the zeolite to regenerate the Lewis acid site of the zeolite or when a carbenium ion donates a proton to regenerate the Brønsted site of the zeolite and produce an olefin (Eq. 2.10).

$$(2.10)$$

Catalytic cracking can also generate by-products such as the formation of aromatic compounds derived from benzene (Figure 2.7).[56]

Simple aromatic compounds in the presence of long-chain hydrocarbons can give rise to polyaromatic compounds (Figure 2.8).[57]

Since the 1980s, ZSM-5 has been added to zeolite Y catalyst to improve olefin production. Chemically, ZSM-5 rapidly consumes the carbenium ions generated by zeolite Y during primary cracking. However, in the absence of ZSM-5, the carbenium ions participate in the hydrogen transfer reaction mechanism and consequently, the olefin yield decreases. The use of ZSM-5 as an additive to the FCC catalyst results in the formation of a higher amount of propylene, which is a precursor of high-value polymers (i.e., poly-acrylates, -ethoxylates, and -ethylene), fibers, and other chemicals (i.e., isopropanol, acetic acid, ethanolamine, and acetaldehyde).[49,55,58,59]

Recently, a first principles-based microkinetic analysis was carried out that allowed a rational tuning of the monomolecular, bimolecular, and aromatization mechanisms in the catalytic pyrolysis of hexane over ZSM-5.[60]

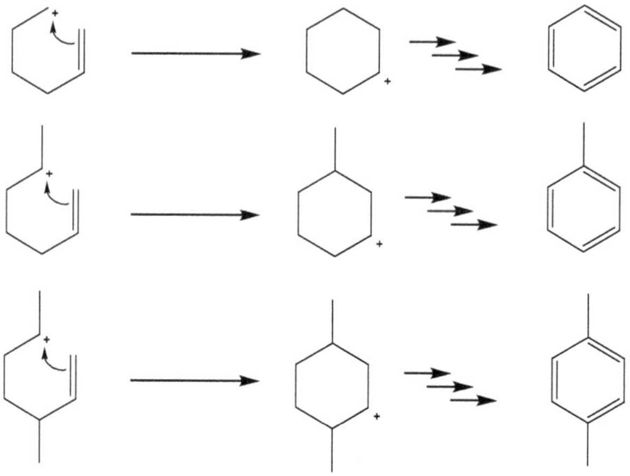

FIGURE 2.7   Formation of aromatic compounds by olefin cyclization.

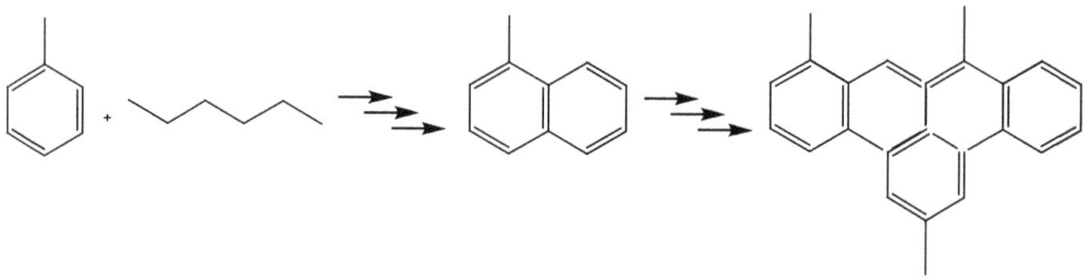

FIGURE 2.8   Formation of polyaromatic compounds.

Catalyst deactivation in the FCC process in the oil industry is a common phenomenon that can occur due to physical causes such as sintering/agglomeration, occlusion, attrition, thermal degradation, and others or due to chemical causes.[61] Some of the causes of deactivation are described below.

Chemical degradation: The reaction of the catalyst with some compounds can cause the destruction or loss of its activity. For example, alkali metals neutralize the acid sites of the catalyst decreasing its activity, zeolite dealumination caused by steam, and vanadium destroying the zeolite structure.[62–64]

Catalyst poisoning: Catalysts can be poisoned by the irreversible adsorption of impurities on their active sites present in the feed stream, such as heavy metals like nickel, vanadium, iron, and sodium, organic compounds containing oxygen, sulfur, and nitrogen, or coke. These contaminants block the catalyst's active sites and reduce its ability to promote cracking reactions.[65,66]

Thermal Deactivation: High temperatures in the FCC reactor can cause thermal decomposition of the catalyst's active components. This process can lead to the formation of carbonaceous species, which coat the catalyst's surface and reduce its catalytic activity. This phenomenon is known as temperature sintering. Thermal degradation of catalysts is an irreversible process that requires catalyst replacement after several periods of use.[67]

Loss of Active Surface by Attrition: Over time, the active surface of the catalyst can suffer from erosion due to mechanical abrasion caused by the circulation of solid particles in the fluidized bed. This reduces the availability of active sites for cracking reactions.[68]

Coke formation: During the cracking process, some carbonaceous products may deposit on the catalyst surface, forming coke, which is a porous solid (H/C = 0.3–1.0) formed by the pyrolysis of coal in closed systems at very high temperatures, such as those that characterize the FCC process. Coke can block the pores of the catalyst and decrease its ability to adsorb and desorb feed molecules, leading to a decrease in catalytic activity. Coke also can block the active sites of the catalyst. Coke can be removed by combustion, so the deactivation of the catalyst caused by coke is a reversible deactivation process.[69]

Catalyst Aging: Over time, catalysts can undergo structural and chemical changes due to prolonged exposure to high temperatures and aggressive process conditions. These changes can affect the catalyst's pore distribution and chemical composition, resulting in a decrease in its activity and selectivity.[70]

ZSM-5 is a very acidic zeolite composed of several pentasil (*eight five-membered rings*) units widely used as an additive in the formulation of catalysts for the FCC process. ZSM-5 as zeolite additive and conventionally Y-zeolite (faujasite) used as FCC catalyst increase the octane number of gasoline and significant increase of propylene and butylenes. ZSM-5 also undergoes significant deactivation and structure destruction by dealumination in the presence of nickel and vanadium.[71]

To mitigate the deactivation of FCC catalysts, strategies such as pretreatment of the feed stream to remove contaminants, periodic regeneration of the catalyst by burning accumulated coke, and development of catalysts that are more resistant to poisoning and thermal deactivation, use of additives and more recently machine learning-assisted catalytic design are employed.[72] In addition, monitoring and controlling process conditions can help minimize deactivation and extend catalyst life.

Even though FCC is a mature technology, the study of the catalyst deactivation processes, and their improvements are still under investigation. Deactivation is a very complex phenomenon, as it depends on many process variables such as catalyst composition (zeolite, matrix, binders, and additives), feed composition, and operating conditions.

More recent studies on the deactivation of FCC catalysts at the single particle level have led to a more detailed understanding of the deactivation mechanisms.[73]

A recent study on the thermal and catalytic cracking of three whole crude oils under high severity conditions (600–650°C) using an equilibrium catalyst (E-Cat) and ZSM-5 as the additive showed that the catalytic cracking of crude oils gave naphtha yields of 23.0–34.2 wt. % and more than 40% of C2–C4 olefins. Thermal cracking produced half as many light olefins as catalytic cracking. The naphtha fraction produced under E-Cat/ZSM-5 cracking contained the highest number of aromatics (74 wt.%) and the lowest number of olefins (1 wt.%).[74]

2.6 DEACTIVATION OF FCC CATALYSTS

FCC is a key process for converting heavy petroleum products into lighter, more valuable products such as gasoline and diesel fuel. However, catalyst deactivation can reduce the process's efficiency and productivity. Due to the rapid deactivation, the FCC catalyst continuously moves between the FCC reactor and the regenerator to recover activity. The coke is converted into volatile compounds such as CO_2, CO, H_2O, SO_x, and NO_x in the regenerator.[75]

2.7 FCC PROCESS REACTION KINETICS

Reaction kinetics in the FCC process is fundamental to understanding how system conditions relate to species consumption and formation per unit time. This information is obtained from kinetic parameters and models describing the reactions occurring in catalytic cracking. 3D numerical models are used to study the interaction of hydrodynamics, heat transfer, and catalytic cracking reactions in commercial FCC reactors.[76]

There are several approaches to model catalytic cracking reactions, from fundamental models to more complex models at the molecular level. Some methods for determining reaction kinetics include the power law and the Langmuir–Hinshelwood–Hougen–Watson model. Kinetic models involve the synthesis of feedstock matrices and can be complex, such as the hybrid structural unit and bond-electron matrix (SUBEM) scheme at the molecular level or the structure-oriented lumping (SOL) method for correlating temperature and feedback in the reaction network with yield and product quality.

The development of kinetic models at the molecular level facilitates the optimization of reaction processes to obtain high-performance gasoline. Although these models are often complex, they aid in the performance and optimization of the FCC process.[77]

Innovations for the FCC process and catalyst continue to adapt to existing feedstock and the generation and optimization of new catalytic cracking routes. They also respond to new needs in reaction products, improving the yield of naphtha components and aromatics, reducing gaseous fractions, and extending catalyst life.

The exploration of new sources for obtaining fuels and petrochemical products, such as ethylene and propylene, with cheaper raw materials and low CO_2 emissions, together with the growing demand for fuels, is driving the search for biomass-derived fuels. All this is taking place in the context of reducing fossil hydrocarbon sources.

In summary, the FCC process continues to evolve and adapt to today's challenges and demands, and there is still much to explore and discover in this field.

2.8 CONCLUSIONS

Light olefins represent one of the fundamental elements in the petrochemical industry. Its major production comes from the FCC process. FCC units consist, in general terms, of reactor, regenerator, and catalyst transport lines to produce value-added chemical products. This process employs different types of zeolites, those modified with metals such as rare earth oxides the most widely used, showing remarkable activity in the catalytic cracking of naphtha, with high selectivity toward light olefins.

The distribution of desired products is closely linked to the chain propagation mechanism and the number of propagation events per initiation step, variables that can be controlled by the reaction conditions and the characteristics of the catalyst used. Although the cracking mechanism of olefins and alkyl aromatics is well established, it still requires further understanding, especially regarding the activation pathway, depending largely on the type of raw materials, operating conditions, and catalyst regeneration.

REFERENCES

(1) Velázquez, H. D.; Cerón-Camacho, R.; Mosqueira-Mondragón, M. L.; Hernández-Cortez, J. G.; Montoya De La Fuente, J. A.; Hernández-Pichardo, M. L.; Beltrán-Oviedo, T. A.; Martínez-Palou, R. Recent Progress on Catalyst Technologies for High Quality Gasoline Production. *Catal. Rev.* 2023, *65* (4), 1079–1299. https://doi.org/10.1080/01614940.2021.2003084

(2) Pinheiro, C. I. C.; Fernandes, J. L.; Domingues, L.; Chambel, A. J. S.; Graça, I.; Oliveira, N. M. C.; Cerqueira, H. S.; Ribeiro, F. R. Fluid Catalytic Cracking (FCC) Process Modeling, Simulation, and Control. *Ind. Eng. Chem. Res.* 2012, *51* (1), 1–29. https://doi.org/10.1021/ie200743c

(3) Oloruntoba, A.; Zhang, Y.; Hsu, C. S. State-of-the-Art Review of Fluid Catalytic Cracking (FCC) Catalyst Regeneration Intensification Technologies. *Energies* 2022, *15* (6), 2061. https://doi.org/10.3390/en15062061

(4) Gandhi, D.; Bandyopadhyay, R.; Soni, B. Naturally Occurring Bentonite Clay: Structural Augmentation, Characterization and Application as Catalyst. *Mater. Today Proc.* 2022, *57*, 194–201. https://doi.org/10.1016/j.matpr.2022.02.346

(5) Khande, A. R.; Dasila, P. K.; Majumder, S.; Maity, P.; Thota, C. Recent Developments in FCC Process and Catalysts. In *Catalysis for Clean Energy and Environmental Sustainability*; Pant, K. K., Gupta, S. K., Ahmad, E., Eds.; Springer International Publishing: Cham, 2021; pp 65–108. https://doi.org/10.1007/978-3-030-65021-6_3

(6) Ferreira, J. M. M.; Sousa-Aguiar, E. F.; Aranda, D. A. G. FCC Catalyst Accessibility—A Review. *Catalysts* 2023, *13* (4), 784. https://doi.org/10.3390/catal13040784

(7) Akah, A.; Al-Ghrami, M. Maximizing Propylene Production via FCC Technology. *Appl. Petrochem. Res.* 2015, *5* (4), 377–392. https://doi.org/10.1007/s13203-015-0104-3

(8) Lin, Y.-H. Production of Valuable Hydrocarbons by Catalytic Degradation of a Mixture of Post-Consumer Plastic Waste in a Fluidized-Bed Reactor. *Polym. Degrad. Stab.* 2009, *94* (11), 1924–1931. https://doi.org/10.1016/j.polymdegradstab.2009.08.004

(9) Chen, Y.-M. Recent Advances in FCC Technology. *Powder Technol.* 2006, *163* (1–2), 2–8. https://doi.org/10.1016/j.powtec.2006.01.001

(10) Ebrahimi, S.; Moghaddas, J. S.; Aghjeh, M. K. R. Study on Thermal Cracking Behavior of Petroleum Residue. *Fuel* 2008, *87* (8–9), 1623–1627. https://doi.org/10.1016/j.fuel.2007.08.015

(11) Bai, P.; Etim, U. J.; Yan, Z.; Mintova, S.; Zhang, Z.; Zhong, Z.; Gao, X. Fluid Catalytic Cracking Technology: Current Status and Recent Discoveries on Catalyst Contamination. *Catal. Rev.* 2019, *61* (3), 333–405. https://doi.org/10.1080/01614940.2018.1549011

(12) Corma, A.; Sauvanaud, L.; Mathieu, Y.; Al-Bogami, S.; Bourane, A.; Al-Ghrami, M. Direct Crude Oil Cracking for Producing Chemicals: Thermal Cracking Modeling. *Fuel* 2018, *211*, 726–736. https://doi.org/10.1016/j.fuel.2017.09.099

(13) Sadeghbeigi, R. *Fluid Catalytic Cracking Handbook: An Expert Guide to the Practical Operation, Design, and Optimization of FCC Units*, Fourth edition.; Butterworth-Heinemann, an imprint of Elsevier: Amsterdam Oxford Cambridge, MA, 2020.

(14) García-Martínez, J.; Johnson, M.; Valla, J.; Li, K.; Ying, J. Y. Mesostructured Zeolite Y—High Hydrothermal Stability and Superior FCC Catalytic Performance. *Catal. Sci. Technol.* 2012, *2* (5), 987. https://doi.org/10.1039/c2cy00309k

(15) Martínez, C.; Vidal-Moya, A.; Yilmaz, B.; Kelkar, C.; Corma, A. Minimizing Rare Earth Content of FCC Catalysts: Understanding the Fundamentals on Combined P-La Stabilization. *Catal. Today* 2023, *418*, 114123. https://doi.org/10.1016/j.cattod.2023.114123

(16) Alotaibi, F. M.; González-Cortés, S.; Alotibi, M. F.; Xiao, T.; Al-Megren, H.; Yang, G.; Edwards, P. P. Enhancing the Production of Light Olefins from Heavy Crude Oils: Turning Challenges into Opportunities. *Catal. Today* 2018, *317*, 86–98. https://doi.org/10.1016/j.cattod.2018.02.018

(17) Bryden, K.; Singh, U.; Berg, M.; Brandt, S.; Schiller, R.; Cheng, W. Fluid Catalytic Cracking (FCC): Catalysts and Additives. In *Kirk-Othmer Encyclopedia of Chemical Technology*; Kirk-Othmer, Ed.; Wiley, 2015; pp 1–37. https://doi.org/10.1002/0471238961.fluidnee.a01.pub2

(18) Corma, A.; Wojciechowski, B. W. The Chemistry of Catalytic Cracking. *Catal. Rev.* 1985, *27* (1), 29–150. https://doi.org/10.1080/01614948509342358

(19) Salahudeen, N.; Ahmed, A. S.; Al-Muhtaseb, A. H.; Dauda, M.; Jibril, B. Y.; Viswanadham, N.; Saxena, S. K. Synthesis of RE Y Zeolite for Formulation of FCC Catalyst and the Catalytic Performance in Cracking of N-Hexadecane. *Res. Chem. Intermed.* 2017, *43* (1), 467–479. https://doi.org/10.1007/s11164-016-2635-3

(20) Gao, X.; Qin, Z.; Wang, B.; Zhao, X.; Li, J.; Zhao, H.; Liu, H.; Shen, B. High Silica REHY Zeolite with Low Rare Earth Loading as High-Performance Catalyst for Heavy Oil Conversion. *Appl. Catal. Gen.* 2012, *413–414*, 254–260. https://doi.org/10.1016/j.apcata.2011.11.015

(21) Ostroumova, V. A.; Maksimov, A. L. MWW-Type Zeolites: MCM-22, MCM-36, MCM-49, and MCM-56 (A Review). *Pet. Chem.* 2019, *59* (8), 788–801. https://doi.org/10.1134/S0965544119080140

(22) Ibarra, Á.; Hita, I.; Azkoiti, M. J.; Arandes, J. M.; Bilbao, J. Catalytic Cracking of Raw Bio-Oil under FCC Unit Conditions over Different Zeolite-Based Catalysts. *J. Ind. Eng. Chem.* 2019, *78*, 372–382. https://doi.org/10.1016/j.jiec.2019.05.032

(23) Zheng, Q.; Huo, L.; Li, H.; Mi, S.; Li, X.; Zhu, X.; Deng, X.; Shen, B. Exploring Structural Features of USY Zeolite in the Catalytic Cracking of Jatropha Curcas L. Seed Oil towards Higher Gasoline/Diesel Yield and Lower CO_2 Emission. *Fuel* 2017, *202*, 563–571. https://doi.org/10.1016/j.fuel.2017.04.073

(24) Sousa-Aguiar, E. F.; Trigueiro, F. E.; Zotin, F. M. Z. The Role of Rare Earth Elements in Zeolites and Cracking Catalysts. *Catal. Today* 2013, *218–219*, 115–122. https://doi.org/10.1016/j.cattod.2013.06.021

(25) Ibarra, Á.; Hita, I.; Azkoiti, M. J.; Arandes, J. M.; Bilbao, J. Catalytic Cracking of Raw Bio-Oil under FCC Unit Conditions over Different Zeolite-Based Catalysts. *J. Ind. Eng. Chem.* 2019, *78*, 372–382. https://doi.org/10.1016/j.jiec.2019.05.032

(26) Nazarova, G.; Ivashkina, E.; Ivanchina, E.; Oreshina, A.; Vymyatnin, E. A Predictive Model of Catalytic Cracking: Feedstock-Induced Changes in Gasoline and Gas Composition. *Fuel Process. Technol.* 2021, *217*, 106720. https://doi.org/10.1016/j.fuproc.2020.106720

(27) Zhao, Y.; Liu, J.; Xiong, G.; Guo, H. Enhancing Hydrothermal Stability of Nano-Sized HZSM-5 Zeolite by Phosphorus Modification for Olefin Catalytic Cracking of Full-Range FCC Gasoline. *Chin. J. Catal.* 2017, *38* (1), 138–145. https://doi.org/10.1016/S1872-2067(16)62579-2

(28) Vu, X. H.; Armbruster, U. Catalytic Cracking of Triglycerides over Micro/Mesoporous Zeolitic Composites Prepared from ZSM-5 Precursors with Varying Aluminum Contents. *React. Kinet. Mech. Catal.* 2018, *125* (1), 381–394. https://doi.org/10.1007/s11144-018-1415-z

(29) Gurdeep Singh, H. K.; Yusup, S.; Quitain, A. T.; Kida, T.; Sasaki, M.; Cheah, K. W.; Ameen, M. Production of Gasoline Range Hydrocarbons from Catalytic Cracking of Linoleic Acid over Various Acidic Zeolite Catalysts. *Environ. Sci. Pollut. Res.* 2019, *26* (33), 34039–34046. https://doi.org/10.1007/s11356-018-3223-4

(30) Yu, Q.; Sun, H.; Sun, H.; Li, L.; Zhu, X.; Ren, S.; Guo, Q.; Shen, B. Highly Mesoporous IM-5 Zeolite Prepared by Alkaline Treatment and Its Catalytic Cracking Performance. *Microporous Mesoporous Mater.* 2019, *273*, 297–306. https://doi.org/10.1016/j.micromeso.2018.08.016

(31) Gandhi, D.; Bandyopadhyay, R.; Soni, B. Naturally Occurring Bentonite Clay: Structural Augmentation, Characterization and Application as Catalyst. *Mater. Today Proc.* 2022, *57*, 194–201. https://doi.org/10.1016/j.matpr.2022.02.346

(32) Ahmad, M.; Farhana, R.; Raman, A. A. A.; Bhargava, S. K. Synthesis and Activity Evaluation of Heterometallic Nano Oxides Integrated ZSM-5 Catalysts for Palm Oil Cracking to Produce Biogasoline. *Energy Convers. Manag.* 2016, *119*, 352–360. https://doi.org/10.1016/j.enconman.2016.04.069

(33) Hussain, A. I.; Palani, A.; Aitani, A. M.; Čejka, J.; Shamzhy, M.; Kubů, M.; Al-Khattaf, S. S. Catalytic Cracking of Vacuum Gasoil over -SVR, ITH, and MFI Zeolites as FCC Catalyst Additives. *Fuel Process. Technol.* 2017, *161*, 23–32. https://doi.org/10.1016/j.fuproc.2017.01.050

(34) Ancheyta-Juárez, J.; Murillo-Hernández, J. A. A Simple Method for Estimating Gasoline, Gas, and Coke Yields in FCC Processes. *Energy Fuels* 2000, *14* (2), 373–379. https://doi.org/10.1021/ef990140y

(35) De La Puente, G.; Devard, A.; Sedran, U. Conversion of Residual Feedstocks in FCC. Evaluation of Feedstock Reactivity and Product Distributions in the Laboratory. *Energy Fuels* 2007, *21* (6), 3090–3094. https://doi.org/10.1021/ef700313u

(36) Gutiérrez Sama, S.; Barrère-Mangote, C.; Bouyssière, B.; Giusti, P.; Lobinski, R. Recent Trends in Element Speciation Analysis of Crude Oils and Heavy Petroleum Fractions. *TrAC Trends Anal. Chem.* 2018, *104*, 69–76. https://doi.org/10.1016/j.trac.2017.10.014

(37) Li, Y.; Shang, H.; Zhang, Q.; Elabyouki, M.; Zhang, W. Theoretical Study of the Structure and Properties of Ni/V Porphyrins under Microwave Electric Field: A DFT Study. *Fuel* 2020, *278*, 118305. https://doi.org/10.1016/j.fuel.2020.118305

(38) Nazarova, G. Y.; Ivashkina, E. N.; Ivanchina, E. D.; Mezhova, M. Y. A Model of Catalytic Cracking: Catalyst Deactivation Induced by Feedstock and Process Variables. *Catalysts* 2022, *12* (1), 98. https://doi.org/10.3390/catal12010098

(39) Scherzer, J. Octane-Enhancing, Zeolitic FCC Catalysts: Scientific and Technical Aspects. *Catal. Rev.* 1989, *31* (3), 215–354. https://doi.org/10.1080/01614948909349934

(40) O'Connor, P. Chapter 15 Catalytic Cracking: The Future of an Evolving Process. In *Studies in Surface Science and Catalysis*; Elsevier, 2007; Vol. 166, pp 227–251. https://doi.org/10.1016/S0167-2991(07)80198-4

(41) Vogt, E. T. C.; Whiting, G. T.; Dutta Chowdhury, A.; Weckhuysen, B. M. Zeolites and Zeotypes for Oil and Gas Conversion. In *Advances in Catalysis*; Elsevier, 2015; Vol. 58, pp 143–314. https://doi.org/10.1016/bs.acat.2015.10.001

(42) Theologos, K. N.; Lygeros, A. I.; Markatos, N. C. Feedstock Atomization Effects on FCC Riser Reactors Selectivity. *Chem. Eng. Sci.* 1999, *54* (22), 5617–5625. https://doi.org/10.1016/S0009-2509(99)00294-8

(43) Magomedov, R. N.; Popova, A. Z.; Maryutina, T. A.; Kadiev, Kh. M.; Khadzhiev, S. N. Current Status and Prospects of Demetallization of Heavy Petroleum Feedstock (Review). *Pet. Chem.* 2015, *55* (6), 423–443. https://doi.org/10.1134/S0965544115060092

(44) Xu, C.; Gao, J.; Zhao, S.; Lin, S. Correlation between Feedstock SARA Components and FCC Product Yields. *Fuel* 2005, *84* (6), 669–674. https://doi.org/10.1016/j.fuel.2004.08.009

(45) Mallah, M. A.; Changxing, L.; Mallah, M. A.; Noreen, S.; Liu, Y.; Saeed, M.; Xi, H.; Ahmed, B.; Feng, F.; Mirjat, A. A.; Wang, W.; Jabar, A.; Naveed, M.; Li, J.-H.; Zhang, Q. Polycyclic Aromatic Hydrocarbon and Its Effects on Human Health: An Overview. *Chemosphere* 2022, *296*, 133948. https://doi.org/10.1016/j.chemosphere.2022.133948

(46) Sadrameli, S. M. Thermal/Catalytic Cracking of Liquid Hydrocarbons for the Production of Olefins: A State-of-the-Art Review II: Catalytic Cracking Review. *Fuel* 2016, *173*, 285–297. https://doi.org/10.1016/j.fuel.2016.01.047

(47) Kotrel, S.; Knözinger, H.; Gates, B. C. The Haag–Dessau Mechanism of Protolytic Cracking of Alkanes. *Microporous Mesoporous Mater.* 2000, *35–36*, 11–20. https://doi.org/10.1016/S1387-1811(99)00204-8

(48) Matar, S.; Hatch, L. F. Chapter Three - Crude Oil Processing and Production of Hydrocarbon Intermediates. In *Chemistry of Petrochemical Processes* (Second Edition); Matar, S., Hatch, L. F., Eds.; Gulf Professional Publishing: Woburn, 2001; pp 49–110. https://doi.org/10.1016/B978-088415315-3/50004-3

(49) Almuqati, N. S.; Aldawsari, A. M.; Alharbi, K. N.; González-Cortés, S.; Alotibi, M. F.; Alzaidi, F.; Dilworth, J. R.; Edwards, P. P. Catalytic Production of Light Olefins: Perspective and Prospective. *Fuel* 2024, *366*, 131270. https://doi.org/10.1016/j.fuel.2024.131270

(50) Sadrameli, S. M. Thermal/Catalytic Cracking of Liquid Hydrocarbons for the Production of Olefins: A State-of-the-Art Review II: Catalytic Cracking Review. *Fuel* 2016, *173*, 285–297. https://doi.org/10.1016/j.fuel.2016.01.047

(51) Vogt, E. T.; Weckhuysen, B. Fluid Catalytic Cracking: Recent Developments on the Grand Old Lady of Zeolite Catalysis. *Chem. Soc. Rev.* 2015, *44* (20), 7342–7370. https://doi.org/10.1039/C5CS00376H

(52) Haag, W. O.; Dessau, R. M.; Lago, R. M. Kinetics and Mechanism of Paraffin Cracking with Zeolite Catalysts. In *Studies in Surface Science and Catalysis*; Lnui, T., Namba, S., Tatsumi, T., Eds.; Chemistry of Microporous Crystals; Elsevier, 1991; Vol. 60, pp 255–265. https://doi.org/10.1016/S0167-2991(08)61903-5

(53) Corma, A.; Orchillés, A. V. Current Views on the Mechanism of Catalytic Cracking. *Microporous Mesoporous Mater.* 2000, *35–36*, 21–30. https://doi.org/10.1016/S1387-1811(99)00205-X

(54) Abrevaya, H. Cracking of Naphtha Range Alkanes and Naphthenes over Zeolites. In *Studies in Surface Science and Catalysis*; Xu, R., Gao, Z., Chen, J., Yan, W., Eds.; From Zeolites to Porous MOF Materials - The 40th Anniversary of International Zeolite Conference; Elsevier, 2007; Vol. 170, pp 1244–1251. https://doi.org/10.1016/S0167-2991(07)80984-0

(55) Alotaibi, F. M.; González-Cortés, S.; Alotibi, M. F.; Xiao, T.; Al-Megren, H.; Yang, G.; Edwards, P. P. Enhancing the Production of Light Olefins from Heavy Crude Oils: Turning Challenges into Opportunities. *Catal. Today* 2018, *317*, 86–98. https://doi.org/10.1016/j.cattod.2018.02.018

(56) Oloruntoba, A.; Zhang, Y.; Hsu, C. State-of-the-Art Review of Fluid Catalytic Cracking (FCC) Catalyst Regeneration Intensification Technologies. *Energies* 2022, *15*, 2061. https://doi.org/10.3390/en15062061

(57) Kang, X.; Guo, X.; You, H. An Introduction to the Lump Kinetics Model and Reaction Mechanism of FCC Gasoline. *Energy Sources Part Recovery Util. Environ. Eff.* 2013, *35* (20), 1921–1928. https://doi.org/10.1080/15567036.2010.531508

(58) Alotibi, M. F.; Alshammari, B. A.; Alotaibi, M. H.; Alotaibi, F. M.; Alshihri, S.; Navarro, R. M.; Fierro, J. L. G. ZSM-5 Zeolite Based Additive in FCC Process: A Review on Modifications for Improving Propylene Production. *Catal. Surv. Asia* 2020, *24* (1), 1–10. https://doi.org/10.1007/s10563-019-09285-1

(59) Gholami, Z.; Gholami, F.; Tišler, Z.; Tomas, M.; Vakili, M. A Review on Production of Light Olefins via Fluid Catalytic Cracking. *Energies* 2021, *14* (4), 1089. https://doi.org/10.3390/en14041089

(60) Chen, D.; Liu, D.; He, H.; Zhao, L.; Gao, J.; Xu, C. Rational Tuning of Monomolecular, Bimolecular and Aromatization Pathways in the Catalytic Pyrolysis of Hexane on ZSM-5 from a First-Principles-Based Microkinetics Analysis. *Fuel* 2024, *366*, 131368. https://doi.org/10.1016/j.fuel.2024.131368

(61) Adanenche, D. E.; Aliyu, A.; Atta, A. Y.; El-Yakubu, B. J. Residue Fluid Catalytic Cracking: A Review on the Mitigation Strategies of Metal Poisoning of RFCC Catalyst Using Metal Passivators/Traps. *Fuel* 2023, *343*, 127894. https://doi.org/10.1016/j.fuel.2023.127894

(62) Etim, U. J.; Bai, P.; Liu, X.; Subhan, F.; Ullah, R.; Yan, Z. Vanadium and Nickel Deposition on FCC Catalyst: Influence of Residual Catalyst Acidity on Catalytic Products. *Microporous Mesoporous Mater.* 2019, *273*, 276–285. https://doi.org/10.1016/j.micromeso.2018.07.011

(63) Psarras, A. C.; Iliopoulou, E. F.; Kostaras, K.; Lappas, A. A.; Pouwels, C. Investigation of Advanced Laboratory Deactivation Techniques of FCC Catalysts via FTIR Acidity Studies. *Microporous Mesoporous Mater.* 2009, *120* (1), 141–146. https://doi.org/10.1016/j.micromeso.2008.09.014

(64) Alabdullah, M. A.; Shoinkhorova, T.; Dikhtiarenko, A.; Ould-Chikh, S.; Rodriguez-Gomez, A.; Chung, S.; Alahmadi, A. O.; Hita, I.; Pairis, S.; Hazemann, J.; Castaño, P.; Ruiz-Martinez, J.; Osorio, I. M.; Almajnouni, K.; Xu, W.; Gascon, J. Understanding Catalyst Deactivation during the Direct Cracking of Crude Oil. *Catal. Sci. Technol.* 2022, *12* (18), 5657–5670. https://doi.org/10.1039/D2CY01125E

(65) Souza, N. L. A.; Tkach, I.; Morgado, E.; Krambrock, K. Vanadium Poisoning of FCC Catalysts: A Quantitative Analysis of Impregnated and Real Equilibrium Catalysts. *Appl. Catal. Gen.* 2018, *560*, 206–214. https://doi.org/10.1016/j.apcata.2018.05.003

(66) Tangstad, E.; Andersen, A.; Myhrvold, E. M.; Myrstad, T. Catalytic Behaviour of Nickel and Iron Metal Contaminants of an FCC Catalyst after Oxidative and Reductive Thermal Treatments. *Appl. Catal. Gen.* 2008, *346* (1), 194–199. https://doi.org/10.1016/j.apcata.2008.05.022

(67) Cerqueira, H. S.; Caeiro, G.; Costa, L.; Ramôa Ribeiro, F. Deactivation of FCC Catalysts. *J. Mol. Catal. Chem.* 2008, *292* (1), 1–13. https://doi.org/10.1016/j.molcata.2008.06.014

(68) Froment, G. F. Modeling of Catalyst Deactivation. *Appl. Catal. Gen.* 2001, *212* (1), 117–128. https://doi.org/10.1016/S0926-860X(00)00850-4

(69) Zhou, J.; Zhao, J.; Zhang, J.; Zhang, T.; Ye, M.; Liu, Z. Regeneration of Catalysts Deactivated by Coke Deposition: A Review. *Chin. J. Catal.* 2020, *41* (7), 1048–1061. https://doi.org/10.1016/S1872-2067(20)63552-5

(70) Stratiev, D.; Shishkova, I.; Ivanov, M.; Chavdarov, I.; Yordanov, D. Dependence of Fluid Catalytic Cracking Unit Performance on H-Oil Severity, Catalyst Activity, and Coke Selectivity. *Chem. Eng. Technol.* 2020, 43 (11), 2266–2276. https://doi.org/10.1002/ceat.202000071.

(71) Gusev, A. A.; Psarras, A. C.; Triantafyllidis, K. S.; Lappas, A. A.; Diddams, P. A.; Vasalos, I. A. ZSM-5 Additive Deactivation with Nickel and Vanadium Metals in the Fluid Catalytic Cracking (FCC) Process. *Ind. Eng. Chem. Res.* 2020, *59* (6), 2631–2641. https://doi.org/10.1021/acs.iecr.9b04819

(72) Bian, J.; Wang, B.; Niu, X.; Zhao, H.; Ling, H.; Ju, F. Migration and Emission Characteristics of Metal Pollutants in Fluid Catalytic Cracking (FCC) Process. *J. Hazard. Mater.* 2024, *462*, 132778. https://doi.org/10.1016/j.jhazmat.2023.132778

(73) Bai, P.; Etim, U. J.; Yan, Z.; Mintova, S.; Zhang, Z.; Zhong, Z.; Gao, X. Fluid Catalytic Cracking Technology: Current Status and Recent Discoveries on Catalyst Contamination. *Catal. Rev.* 2019, *61* (3), 333–405. https://doi.org/10.1080/01614940.2018.1549011

(74) Al-Absi, A. A.; Aitani, A. M.; Al-Khattaf, S. S. Thermal and Catalytic Cracking of Whole Crude Oils at High Severity. *J. Anal. Appl. Pyrolysis* 2020, *145*, 104705. https://doi.org/10.1016/j.jaap.2019.104705

(75) Khande, A. R.; Dasila, P. K.; Majumder, S.; Maity, P.; Thota, C. Recent Developments in FCC Process and Catalysts. In: *Catalysis for Clean Energy and Environmental Sustainability: Petrochemicals and Refining Processes - Volume 2*; Pant, K. K., Gupta, S. K., Ahmad, E., Eds.; Springer International Publishing: Cham, 2021. https://doi.org/10.1007/978-3-030-65021-6

(76) Rodríguez-Fragoso, M.; Zavala-Salazar, S.; Moreno-Montiel, M.; Bouchot, C.; Elizalde-Solis, O.; Ramírez-Jiménez, E. A Kinetic Model for the FCC Process as Function of the Feedstock Composition. *Chem. Eng. J.* 2023, *474*, 145489. https://doi.org/10.1016/j.cej.2023.145489

(77) Chen, Z.; Wang, G.; Zhao, S.; Zhang, L. Prediction of Molecular Distribution and Temperature Profile of FCC Process through Molecular-Level Kinetic Modeling. *Chem. Eng. Sci.* 2022, *264*, 118189. https://doi.org/10.1016/j.ces.2022.118189

3 Catalytic Reforming of Naphtha

Nohra Violeta Gallardo Rivas
Technological Institute of Ciudad Madero, México

Sandra Irene Martínez Torres
Technological Institute of Ciudad Madero, México

Armando Balboa Palomino
Technological Institute of Ciudad Madero, México

Ulises Páramo García
Technological Institute of Ciudad Madero, México

3.1 INTRODUCTION

For more than 70 years, catalytic reforming of naphtha has played a very important role in the petroleum refining industry, the objective of which is the production of gasoline with a high-octane number that can be commercialized. This process is essential in all the petrochemical plants around the world, which consists of the transformation of low-octane naphtha (C_6–C_{10} hydrocarbons) into higher octane products for gasoline blending and contains aromatic compounds for the production of petrochemicals. Hydrogen gas with high purity is also obtained as a subproduct, the demand for which is increasing due to the need for hydrotreatment processes.

Several reactions occur in the reforming catalytic process in order to increase the gasoline octane number. They include (1) dehydrogenation of naphthene compounds, (2) dehydrogenation of kerosene compounds, (3) isomerization of normal kerosenes, and (4) hydrocracking reactions of kerosenes converting them into lower molecular weight kerosenes and hydrodisalkylation of aromatics. Dehydrogenation, dehydrocyclization, and isomerization reactions are desired because they control the octane number and hydrogen purity; on the other hand, hydrocracking reactions affect the molecules of paraffin, causing the molecules to break down and produce light low-octane gases (LPG).

The optimal operating conditions for carrying out catalytic reforming to produce high-octane aromatic compounds are high temperatures, low pressures, and low hydrogen/feed ratios. The hydrogen pressure must be high enough to avoid catalyst deactivation at the surface.[1,2]

DOI: 10.1201/9781003517283-3

Catalytic reforming is one of the main units to produce gasoline in refineries. This unit can produce around 37% (weight) of the total production of gasoline. Other units like fluidic catalytic cracking (FCC) and the alkylation and isomerization units also contribute to this gasoline generation.[3]

The product with the highest quality from the crude distillation unit is hydrotreated to remove sulfur, nitrogen, and oxygen, which can deactivate the catalyst used in reforming. The hydrotreated naphtha is fractionated into light naphtha that is mainly constituted of C_5–C_6 hydrocarbons, and into heavy naphtha, constituted of C_7–C_{10} hydrocarbons. It is very important to remove C_6-type molecules from the reformer feed; otherwise, benzene will be formed (undesired subproduct). Light naphtha is isomerized, and it can be fractionated if it is introduced into the reformer.[4]

The role of the heavy naphtha reformer in a refinery is shown in Figure 3.1. The produced hydrogen in this system can be recycled through the naphtha hydrotreating process, and the remainder is sent to other hydrogen-demanding units.[5]

In Figure 3.2, the feed, operation conditions of the reforming process, and the obtained products are presented in a schematic form. The process can be operated in two ways: at high demand to primarily produce aromatic compounds (80–90% in volume) and at medium demand to produce high-octane gasoline (70% aromatics content).

The feeds are characterized by the Watson factor (K), used for the determination of the hydrocarbon's chemical nature (Eq. (3.1)). This factor is related to the density (H/C ratio; ASTM D86 or ASTM D-1160) and the boiling point (number of carbon atoms; ASTM D-1298).[6]

$$K = \frac{T_b^{1/3}}{\gamma} \tag{3.1}$$

where K is the Watson factor; Tb, is the boiling point for pure hydrocarbons (°R), and γ is the specific gravity. For paraffinic components (P), K handles values between

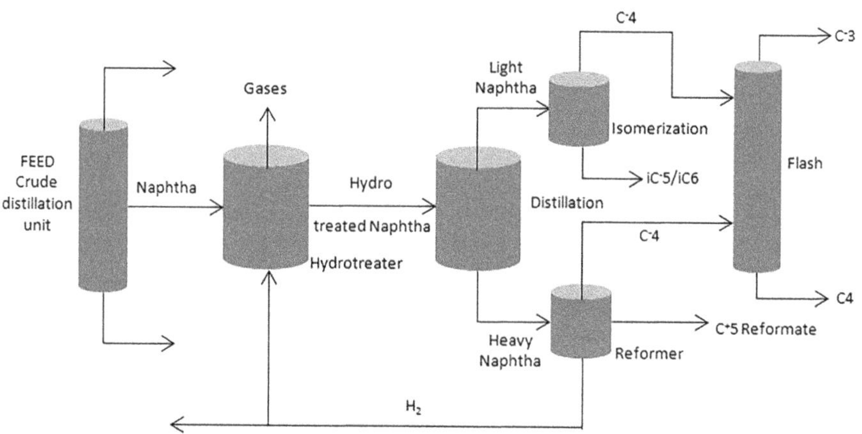

FIGURE 3.1   Reformer in the refinery.

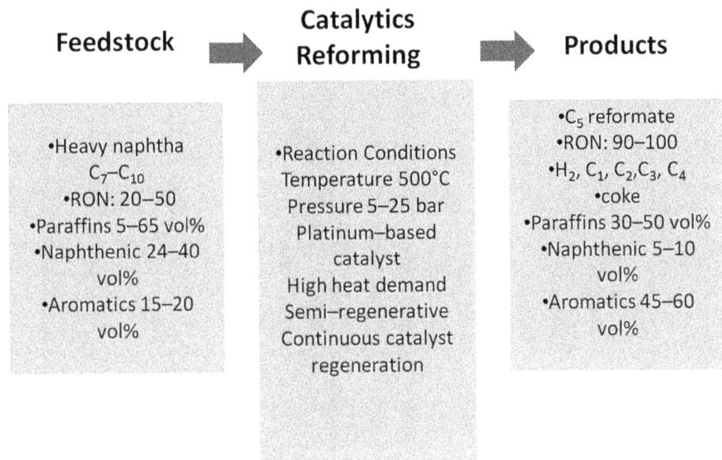

FIGURE 3.2 Conditions of the reforming process.

12.5 and 13.5; for naphthenics (N), between 11.0 and 12.5; and for aromatics (A), between 8.5 and 11.0.

The principal feed comes from hydrotreated heavy naphtha, and a part of that feed comes from hydrotreated coker naphtha.[7] Another key concept in the catalytic reforming process is the research octane number (RON), defined as the volumetric percentage of isooctane in a mixture of isooctane and n-heptane that hits with a certain intensity while the fuel is tested. A list of the RONs of pure hydrocarbons is given in Table 3.1.[8–12]

It can be seen in Table 3.1 that the RONs of paraffins, isoparaffins, and naphthenes decrease as the number of carbon molecules increases. Aromatic compounds have the opposite tendency.[13]

3.2 PRINCIPAL REACTIONS OF CATALYTIC REFORMING

Similar to thermal reforming, catalytic reforming converts low-octane gasoline into high-octane gasoline (reformate). While thermal reforming could produce reformates with RONs ranging from 65 to 80, depending on performance, catalytic reforming produces reformates with RONs in the range of 90 to 95.

Catalytic reforming is performed in the presence of hydrogen over hydrogenation–dehydrogenation catalysts, which can be supported on alumina or silica-alumina. Depending on the catalyst, a defined sequence of reactions involving structural changes in the feedstock takes place. This more modern concept actually made thermal reforming obsolete.[14]

The catalytic reformer feeds are saturated materials (i.e., nonolefinic); in several cases, this feed could be first distillation naphtha, but other by-products of the low-octane naphtha (e.g., coker naphtha) can be processed after treatment to eliminate olefins and other pollutants. The hydrocracking naphtha that contains substantial amounts of naphthenes is also a suitable feed.

TABLE 3.1
Research Octane Numbers of Pure Hydrocarbons (RON)

Paraffins	RON	Olefins	RON	Naphthenes	RON	Aromatic	RON
n-Butane	94	1-Pentene	90	Cyclopentane	>100	Benzene	-
Isobutane	>100	1-Hexene	76.4	Cyclohexane	83	Toluene	>100
n-Pentane	61.8	1-Heptene	54.5	Methylcyclopentane	91.3	o-Xylane	-
2-Methyl-1-butane	92.3	2-Methyl-2-hexene	90.4	Methylcyclohexane	74.8	m-Xylane	>100
n-Hexane	24.8	2,3-Dimethyl-1-pentene	99.3	1,3-Dimethylcyclopentane	80.6	p-Xylane	>100
2-Methyl-1-pentane	73.4	1-Octene	28.7	1,1,3-Trimethylcyclopentane	87.7	Ethylbenzene	>100
2,2-Dimethyl-1-butane	91.8			Ethylcyclohexane	45.6	n-Propylbenzene	>100
n-Heptane	0			Isobutylcyclohexane	33.7	Isopropylbenzene	>100
2-Methylhexane	52					1-Methyl-3-ethylbenzene	>100
2,3-Dimethylpentane	91.1					n-butylbenzene	>100
2,2,3-Trimethylbutane	>100					1-Methyl-3-isopropylbenzene	–
n-Octane	<0					1,2,3,4-Tetramethylbenzene	>100
3,3-Dimethylhexane	75.5						
2,2,4-Trimethylpentane	100						
n-Nonane	<0						
2,2,3,3-Tetramethylpentane	>100						
n-Decane	<0						

3.2.1 DEHYDROGENATION OF NAPHTHENES TO AROMATIC COMPOUNDS

Naphthenic compounds such as cyclohexane, methylcyclohexane, and dimethyl cyclohexane up to C_{10} naphthenes are dehydrogenated to benzene, toluene, and xylenes, respectively. In Figure 3.3, the desired reactions of catalytic reforming are displayed.

The reaction is highly endothermic and looks favored by high temperature and low pressure. Moreover, cyclohexanes with a greater number of carbons produce more aromatic compounds in equilibrium.[15] The kinetics and thermodynamic effect are complex due to the adsorption and desorption of reaction intermediaries which are influenced by the reaction speed. Table 3.2 shows the competitive effects that are influenced by the main parameters governing the dehydrogenation reactions.

Under normal operation conditions, the reforming reaction is very fast and almost complete. It is promoted by the metallic function of the catalyst. From the fact that it produces aromatic compounds with high octane numbers, this reaction is most desirable in petroleum refining.

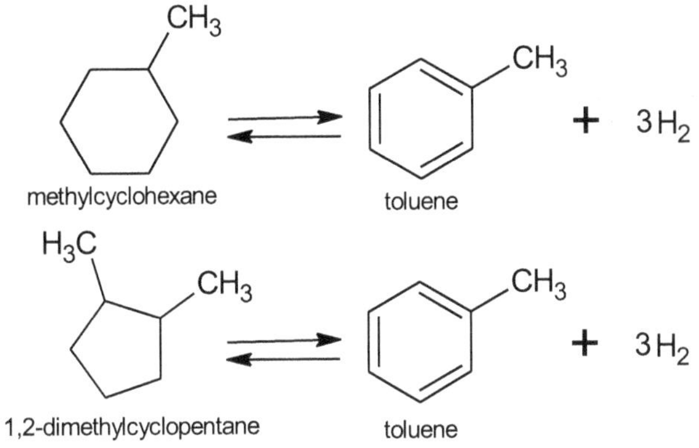

$$n - C_7H_{16} \leftrightarrow n - C_7H_{14} + H_2$$

FIGURE 3.3 Desirable reactions in catalytic reforming.

TABLE 3.2

Impact of Operating Parameters on Naphthene Dehydrogenation

Increasing:	Effect on the Dehydrogenation Yield By:	
	Thermodynamics	**Kinetics**
Pressure	Decreases	Unaffected
Temperature	Increases	Increases
H_2/HC ratio[a]	Slightly decreases	Slightly decreases

[a] Ratio of pure hydrogen(mol) to hydrocarbon feed (mol)

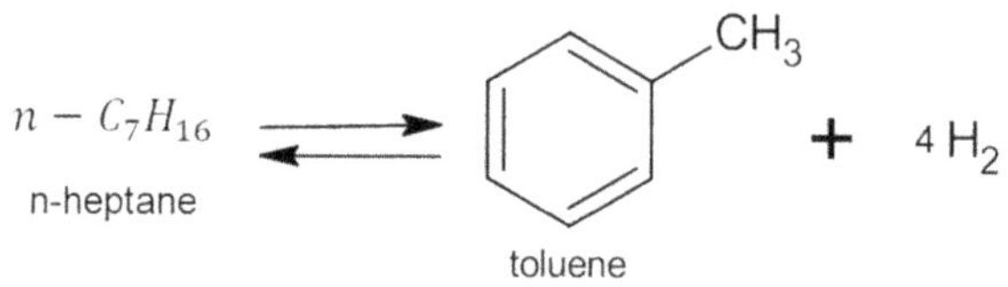

$$n - C_7H_{16} \rightleftharpoons \text{(toluene)} \quad + \quad 4\,H_2$$

n-heptane   toluene

FIGURE 3.4   Dehydrocyclization reaction of paraffins.

3.2.2 DEHYDROCYCLIZATION OF PARAFFINS

The dehydrocyclization of paraffins is another reaction that has great importance in catalytic reforming, and at the same time, is the most complex. This process becomes easier as the molecular weight of paraffin increases, although the paraffins can be hydrocracked simultaneously. Kinetically, the rate of the dehydrocyclization reaction increases at low pressure and high temperature. However, at normal operation conditions, this rate is lower than that of the naphthene dehydrogenation reaction. The reaction is promoted by the acid and metallic function of the catalyst, with the formation of naphthene being the step that controls the reaction rate.[16] The main dehydrocyclization reaction of paraffins is shown in Figure 3.4.

3.2.3 LINEAR ISOMERIZATION OF PARAFFINS

Isomerization rearranges normal low-octane C_5 and C_6 paraffin molecules in light direct distillation naphtha to produce the corresponding isoparaffins C_5 and C_7 of high-octane and, in that way, significantly increases the octane resulting from the naphtha flow (isomerate) to convert it into a valuable blend of gasoline components.

As an additional benefit of the process, isomerization elaborates a product that is virtually sulfur- and benzene-free. Thereby, some refineries have recently added the isomerization capacity as a way to fulfill the strict standards in gasoline production. Figure 3.5 illustrates the main isomerization reaction in this case that of heptane to isoheptane (2-methyylhexane).

The isomerization reaction is partially exothermic and promotes the increase of octane number. From the kinetic perspective, high temperatures favor isomerization and the hydrogen partial pressure does not cause any side effects. These reactions are promoted by the acid and metallic functions of the catalytic support.

3.2.4 HYDROCRACKING

Hydrocracking reactions are the primary source for the production of hydrocarbons up to four carbons (C_1, C_2, C_3, and C_4). This reaction is highly exothermic and

$$n - C_7H_{16} \leftrightarrow iC_7H_{16}$$

n-heptane   isoheptene

FIGURE 3.5   Example of isomerization reaction of paraffins.

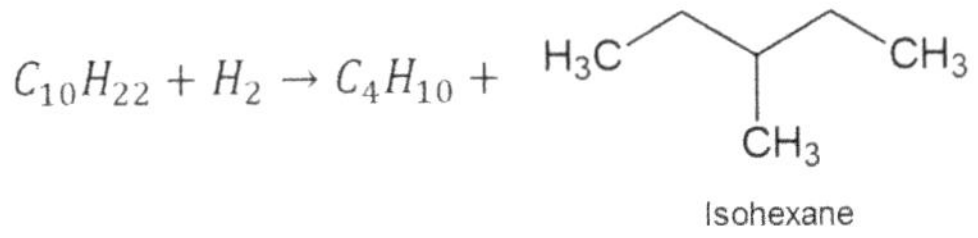

$$C_{10}H_{22} + H_2 \rightarrow C_4H_{10} + \text{(Isohexane)}$$

Isohexane

FIGURE 3.6 Hydrocracking reaction of paraffins.

consumes big amounts of hydrogen. As a result of cracking, low yields are obtained in reforming. Figure 3.6 shows the hydrocracking reaction of paraffins.

Other paraffins could be cracked giving C_1–C_4 molecules as by-products.

Coke can also be deposited during hydrocracking, which leads to catalyst deactivation. In this case, the catalyst must be reactivated by burning the deposited coke. The catalyst is selected to produce a slow hydrocracking reaction. Coke formation is favored at low hydrogen partial pressures.

Hydrocracking is controlled by operating the reaction at low pressure, between 5 and 25 atm (74 and 368 psi), a value not too low for coke deposition and not too high to avoid cracking and loss of reforming performance. In Figure 3.7, the main hydrocracking reaction of aromatics is shown.

A summary of the reactions and interactions of the reformers is given in Figure 3.8.

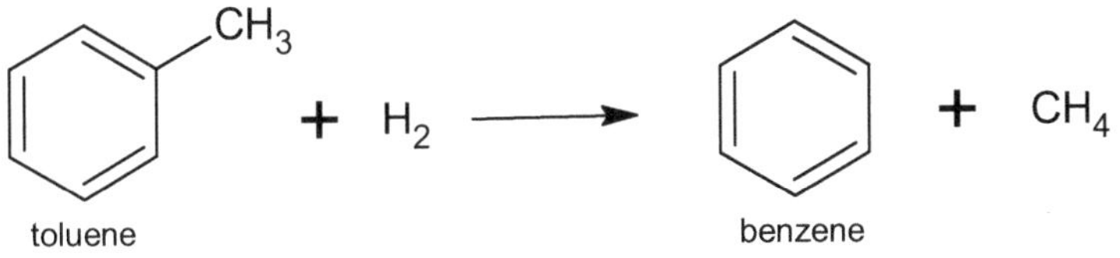

toluene + H_2 ⟶ benzene + CH_4

FIGURE 3.7 Hydrocracking reaction of aromatics.

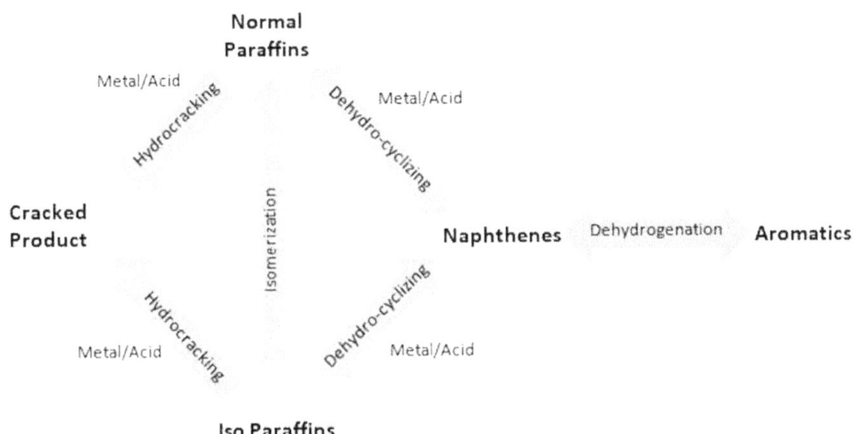

FIGURE 3.8 Network of reforming reaction.

3.3 THERMODYNAMICS OF REFORMING REACTIONS

In light of shifting economic landscapes, there is an inherent necessity for enhancing and optimizing processes. Thermodynamics presents an opportunity to ascertain theoretical energy efficiencies along with the characteristics of process streams under idealized or approximate conditions. This forecasting becomes a valuable asset in crafting new procedures and upgrading current ones. Insights into phase and chemical equilibria hinge on the phase behavior of hydrocarbons and other compounds and necessitate the comprehension of the associated reaction conditions for the reforming and isomerization processes. Such principles of thermodynamics, when applied to phase and chemical equilibria, offer essential understanding for novices in this domain.[17–20]

Over the course of its evolution, catalytic reforming has been continuously refined in response to product specifications, environmental concerns, and economic demands. Originating in the 1920s, it aimed at enhancing thermal naphtha into a high-octane blend via vapor phase processing, utilizing various catalysts and reactors. Its metamorphosis led to platforming a specialized technique producing aromatic-laden reformate with impressive hydrogen efficacy by employing a molten salt/modified alumina catalyst system. Contemporary platforming units churn out such reformate for benzene production and unleaded gasoline. Prompted by the urgency for environmentally benign unleaded gasoline, there has been a pursuit for isomerate that supersedes tetraethyl lead (TEL) as an octane booster.[21–24]

Continual modifications, responsive to the scarcity of high-quality aromatics, and persistent shifts in environmental and economic conditions suggest that the evolution of reforming and isomerization will press on. These alterations have come in response to stringent ecological regulations and the contest for superior transportation fuels. A special focus has been placed on reevaluating established aromatic production methods like catalytic reforming and scrutinizing the entire refining spectrum for newer, more efficient processes. Critical evaluations are underway to identify less rigorous operational conditions and higher hydrogen efficiencies. The goal is twofold: to improve naphthas from catalytic cracking and hydrocracking and to generate high-octane blending stocks compatible with environmental norms and aromatic feedstocks for the chemical industry.[25–27]

3.3.1 REACTIONS

Hydrogenation reactions are the main sources of reformate products and are considered the most important in the process. These are strongly endothermic and require a considerable amount of heat in order to keep the reaction running; for this reason, three reactors are commonly used, where the product leaving each reactor is heated before entering the next reactor.[28]

The reactions are reversible, and equilibrium is established in the function of temperature and pressure. Typically, it is important to calculate the equilibrium conversion for each reaction. Reforming occurs at high temperatures, in the range of 500°C, and low hydrogen pressures are required. The minimum hydrogen partial pressure is determined by the desired quantity of aromatics conversion.[29,30]

Naphtha reforming transforms a petroleum refining by-product of modest value and octane into a superior octane blending agent for crafting high-octane fuel. This entails multiple processes that alter a hydrocarbon's chemical structure, predominantly an increase in aromatic hydrocarbons. This essay emphasizes the fixed bed catalytic reaction for inducing these chemical alterations. Within this context, the predominant reactions are dehydrogenation and dehydroisomerization of paraffins, isomerization of alkanes and paraffins, followed by the cyclization and aromatization processes. These concerning reactions enhance the aromaticity of the hydrocarbons, which is considered crucial for the entire reforming process, leveraging phase diagrams to discern catalysts that maximize reforming outcomes at elevated pressures.[31–35]

3.3.2 EFFECT OF TEMPERATURE ON REACTION EQUILIBRIUM

The propensity for higher yield through endothermic reforming reactions escalates with temperature. However, at temperatures beyond 540°C, catalyst deactivation hastens owing to coke deposition. The equilibrium constant's sensitivity to temperature fluctuations grants insights into the extent to which reactions can be driven when applying the Van't Hoff equation. Additionally, Kirchhoff's equation permits the direct correlation of heat capacity with the reaction's heat, thus facilitating the determination of equilibrium constants across various temperatures. From a thermodynamic perspective, the sign inversion of reaction enthalpy at specific temperatures indicates a pivotal transition from endothermic to exothermic behavior or vice versa.[36–38]

3.3.3 EFFECT OF PRESSURE ON REACTION EQUILIBRIUM

Pivotal for economic viability, increasing pressure to catalyze higher yields of hydrogen from the conversion of naphthenes into paraffins is critical. The formulation

$$\Delta G = \Delta H - T\Delta S, \qquad (3.2)$$

outlines the dependence of enthalpy and entropy alongside ΔG (Eq. (3.2)) on pressure in thermodynamic terms. Reactions like:

$$nC_6H_{12} \rightarrow 2nC_6H_{14} + nH_2$$

show a mole decrease, synonymous with a reduction in ΔS.

This negative entropy shift signals favorable conditions for gas formation, particularly under varied pressures. The absence of comprehensive kinetic and equilibrium data necessitates exploratory assessments of pressure's impact on the ΔG, ΔH, and $-T\Delta S$ variables of select aromatic hydrocarbons within typical naphtha reforming pressure ranges of 1–30 atmospheres.[39,40]

3.3.4 EFFECT OF CATALYST ON REACTION EQUILIBRIUM

In the idealized setting of a catalytic reactor, the catalyst, necessary in only minimal amounts, should theoretically be reusable and not a factor in the completeness of reactions. Platinum on alumina, a common catalyst in naphtha reforming, expedites endothermic reactions by reducing the activation energy required for the reaction, which would otherwise require significant heating of the naphtha stream.

The catalyst's role, ultimately, is to be transparent in the broader reaction, affecting neither the individual heats of reactions nor the equilibrium constant, yet exert a notable influence over the comprehensive heat balance of the naphtha reforming process, impact the thermal profile of successive exothermic and endothermic stages, and suggest an upturn in heat required for the overall reaction.[32,41–43]

3.3.5 INDUSTRIAL APPLICATIONS AND CHALLENGES

Within industrial contexts, the judicious manipulation of thermodynamic properties is pivotal for augmenting outputs of high-value products like gasoline and hydrogen while concurrently curtailing energy demands. These products, which can range from primary targets to intermediary substances such as butanes and more complex hydrocarbons, require redirection through reforming to enhance the overall production of gasoline and hydrogen. The prevailing thermodynamic conditions and specific feedstock composition dictate a state of equilibrium, indicating the proportions of intended yields. Thermodynamic principles suggest that a process will halt upon achieving equilibrium. To boost product yield, shifting said equilibrium toward the product of interest is necessary, which is accomplished by modifying process parameters (T, P, and feed composition) and implementing catalysts. Energy utilization can be quantified by employing Gibbs energy minimization for a single step, with reaction heat and equilibrium constant as known entities and reaction extent as the sole variable.

Moreover, the preeminent aim of industrial naphtha reforming, as posited in this chapter's introduction, is to amplify the yield of gasoline and hydrogen. As understood, favoring the yield of a certain product is contingent upon the strategic shifting of equilibrium. Considering the undesirable by-reactions, like the reforming of butanes and more complex molecules, these can be mitigated or reversed. However, in instances where these reactions occur significantly, they could compete with the primary reforming, negating any increase in desired product yields. To determine such occurrences, hypothetical thermodynamic behavior under surplus pressure is presumed. For a singular reaction, the absence of reaction at insufficient pressure and its reversibility upon lowering the temperature is evidenced by the equation (3.3)

$$\Delta G = RT\ln K, \tag{3.3}$$

which highlights the preference for a high K-value for forward reaction progression.[21,31,32,34,44]

Thermodynamics serves as an influential framework in engineering and refining naphtha reforming methodologies. As addressed previously, the aggregate reaction within naphtha reforming is endothermic, conducted under pressurized conditions conducive to hydrogen generation, yet also bound by thermodynamic limitations. The diminishing equilibrium constants for reforming reactions, associated with increased molecular weight and reaction endothermicity, point out this limitation. Among these, isomerization—paraffin transformation to a higher hydrogenation state—is the favored process, whereas paraffin aromatization is the least favored. Transitioning through the reforming stages, feedstock experiences dehydrogenation and dehydroisomerization, transitioning from paraffins to olefins and from naphthenes to aromatics.[43,45]

Thus, crafting naphtha reforming processes in an era of environmental awareness necessitates constraining hydrogen and by-product formation while bolstering both the caliber and output of the resultant high-octane reformate. Citing a theoretical model from investigative studies, the catalytic efficacy was gauged via a thermodynamic stability constant for the dehydrogenation of a paraffin into an olefin and its reverse reaction. Within this context, a change in the standard state for the proposed dehydrogenation reaction is presupposed, with olefin and paraffin regarded as ideal gases under standard pressure. Exploratory endeavors involved a zinc-on-alumina catalyst, tested under hydrogen and helium at 500°C with C_4–C_7 paraffins, yielding favorable K-values for dehydrogenation, thus buttressing the primary importance of thermodynamic condition settings in catalytic reaction sequences.[46,47]

In chemical process configuration, the pivotal directive is reaching and manipulating the reaction equilibrium, characterized by ΔG. This approach, translated through equation (3.3), yields a slope of $d(n)/dn = K$ with respect to ΔG, where n signifies moles of a substance and r reflects the specific ΔG. By fine-tuning temperature or pressure, K-values can be steered positively to foster desired product synthesis. Such an iterative approach to thermodynamic adjustments encompasses each individual reaction, aiming for a globally optimized system state while respecting catalyst effectiveness and process efficiency.[43,45–48]

Confronted with the task of achieving high octane values during catalytic reforming, two primary thermodynamic hurdles arise. The first relates to the severity of the process, characterized by a swift hydrogen generation rate compared to its consumption in hydrocracking reactions. The intent to maximize aromatic yields while curtailing gaseous paraffin production exacerbates this issue. Under severe reforming conditions, it becomes arduous to limit methane production, favoring instead the production of ethylene and acetylene. Here, the thermodynamics tend toward methane, as the hydrocracking of paraffins and naphthenes, although fast at elevated temperatures, undergo considerable inhibition if the partial pressure of hydrogen decreases. Consequently, the rate of hydrogen generated can considerably outstrip its consumption in ancillary reactions, yielding excess hydrogen for the isomerization, cyclization, and aromatization of paraffins and naphthenes. The second salient thermodynamic challenge relates to the generation of a hydrogen-deficient product. For a prespecified reactor inlet temperature, the product distribution ensues an equilibrium, with the hydrogen deficiency of the product measured against a desired counterpart, gauged by the respective hydrogen contents. Therefore, a higher octane rating correlates with a product H/C ratio inferior to the feed's ratio, dictated by the interplay between hydrogenation and dehydrogenation reactions.[49–51]

3.4 CATALYSTS AND REACTION KINETICS

Catalytic reforming is one of the most important processes in petroleum refining. Catalytic reforming has a major role in motor gasoline production. The process increases the octane number of gasoline and produces hydrogen and aromatic hydrocarbons. Catalytic reforming is a complex system of many interdependent reactions. The process has been the object of many investigations over the years with the aim of establishing the best reaction conditions for the maximum yield and quality of the desired products. Although many studies have been reported, only a portion of the total process has been understood in terms of the reaction kinetics. In order to have a detailed understanding of the reaction processes taking place during catalytic reforming, model-based methods have been used frequently. The models range in sophistication from simple mass balances to complex mathematical treatments. The models often have been directed to design considerations rather than a basic understanding of the reaction mechanism. High aromatic yields are expected in the future and demand for tailor-made benzene will increase. This means that the future processes must be more flexible than the present ones. This underscores the need for a more detailed understanding of the reaction mechanism. On the other hand, it will be more difficult to obtain new environmentally acceptable catalysts. This is because the evaluation of a catalyst has become stricter due to safety and environmental reasons. New regulations concerning the phase out of lead and aromatics and the use of oxygen and hydrogen as fuel cells are expected to tighten. This may result in a decrease in hydrogen consumption in the future. These tougher requirements for future processes have made it necessary to make a new start on the research of catalytic reforming; however, research in this area has been decreasing in recent years.[52–55]

Appropriate kinetic models are essential for the simulation of catalytic reactors and aid in the understanding of the factors that influence catalyst performance. Understanding the chemical kinetics involved in the reactions taking place on the catalyst can lead to the identification of the steps in the reaction sequence that are rate determining. This knowledge can be used to suggest experiments to elucidate the nature of these steps. Kinetic models are used to predict the effect of changes in operating conditions on the severity of the service in the existing unit and to predict the performance of a new unit. If the prediction is that the changes will lead to the deactivation of the catalyst or undesirable side reactions, then the process engineers can use this information to identify alternative conditions or to make changes in the reactor hardware. One of the objectives of such changes might be to obtain a new set of operating conditions that would still achieve the desired outcome in product quality and yield but with a longer on-stream catalyst life.

The ultimate goal would be to use the kinetic model to make a rational choice among various process options. In each case, the process beginning with the suggested changes would be simulated, and the predicted performance compared to the current situation and the desired outcome. By looking at the effect of the proposed changes on the various dependent variables in the kinetic model, it will be possible to assess whether the changes are in fact taking the process in the desired direction. If the prediction is that the changes will not lead to the desired outcome, then alternative changes can be identified and tested. A kinetic model provides a systematic methodology for process optimization.[56–58]

In experimental tests, many have tried to determine the reaction pathway, but different results can be found. This is because the reaction mechanism can still vary under certain conditions. The various methods for kinetic modeling still do not reach a certain agreement, and the experimental data is still insufficient. All of this indicates that kinetic modeling should always be developed alongside experimental tests.

Some authors have studied the deactivation of the catalyst by hydrocarbons. They pointed out that the research has shifted from determining the reaction mechanism and catalyst species to studying the rate of deactivation and catalyst stability.[32,59–62]

This is important because as long as hydrocarbons are still available, the process should be continuously run. The longer the catalyst can activate the reaction, the better the economic factor. However, there are only a few studies on the kinetic modeling of catalytic reforming. According to some authors, until now, kinetic modeling has not been successful because the model is still far from reality and the simulation is difficult to validate. This is because there is no available data on catalyst characteristics and the reaction mechanism of each individual reaction.[44,47,63–66]

The catalysts conventionally used in this process of catalytic reforming are monometallic, bimetallic, and trimetallic, which are supported on alumina, such as platinum (Pt/Al_2O_3), platinum-iridium ($Pt\text{-}Ir/Al_2O_3$), and platinum-iridium-tin ($Pt\text{-}Ir\text{-}Sn/Al_2O_3$), respectively. The catalyst performance could be improved by modifying its stability, selectivity, and activity properties. To achieve high yield and high quality in the reformate, it is necessary to elevate the selectivity of the desired reactions through the equilibrium between the acid and metallic sites.[67,68] Despite the impact of fossil fuels, they are still considered the main source of energy supplies;[69,70] foreseeing this and with the idea of mitigating the environmental impact, work is being done on the generation of laws imposing oil companies for better quality in their products; in the case of gasoline, an increase in octane, which would result in better combustion and less production of pollutants to protect the environment. Having this in mind, it is established that the key point is the refining of hydrocarbons, specifically the catalytic reforming of naphtha, which is used to convert hydrocarbons from low-octane naphtha to more valuable components of high-octane gasoline without changing the boiling point range.[71–73]

The metallic function catalyzes the hydrogenation and dehydrogenation reactions, while the acidic function enhances the isomerization and cyclization reactions. With the purpose of achieving an optimal performance of the catalysts for naphtha reformate, an appropriate balance between the components at stake is needed.[74] Increasing catalyst stability and selectivity, as well as reducing catalyst deactivation, is a vital issue to improve process efficiency and performance. This practice could be achieved by modifying the acid and metal functions by adding components to the support, such as chlorides, which change the strength and quantity of acid sites available.[75] These modifications would raise coking and cracking rates in reactions catalyzed by acids. Although an excessive amount of chlorine would, in hydrocracking reactions, increase carbon deposits, the modification of metal function could be achieved by adding a secondary or ternary metallic component to Pt.[76]

Recently, additional elements have been aggregated to Pt in order to improve the additional properties of the catalyst. Iridium (Ir) is added to enhance activity, Rhenium (Re) is added to operate at lower pressures, and Tin (Sn) is added to improve performance at low pressures.

The use of Pt/Re is nowadays the most common in semi-regenerative processes (SR). Pt/Sn is used in moving bed reactors. The amount of chloride utilized is approximately 1% (weight) of the catalyst, and the amount of Pt is between 0.2 and 0.6% (weight). The potential impurities that could cause deactivation or poisoning of the catalysts include coke, sulfur, nitrogen, metals, and water. Due to these problems, the reformer feed must be severely hydrotreated to eliminate the majority of these, and the reformer must operate at high temperatures and low pressures to minimize coke deposition.[77–79]

The dehydrogenation reactions of paraffins and naphthenes are very fast and normally occur in the first reactor. The isomerization of paraffins and naphthenes are brief, while hydrocracking is slow and takes place in the last reactor. The effect of the operation conditions on the reaction rate and other properties are presented in Table 3.3.[78,79]

TABLE 3.3

Thermodynamic and Kinetic Comparison and Effect of Operating Condition on Main Reactions and Products

Reaction Type	Reactive Rate	Heat Effect	Reaction Equilibrium	Pressure Effect	Temperature Effect	H_2 Production	Reid Vapor Pressure	Product Density	Octane
Naphthene dehydrogenation	Very rapid	Very endo	Yes	−	+	Produce	−	+	+
Naphthene dehydroisomerization	Rapid	Very endo	Yes	−	+	Produce	−	+	+
Paraffin isomerization	Rapid	Slight exo	Yes	None	Slight	No	+	Slight -	+
Paraffin dehydrocyclization	Slow	Very endo	No	−	+	Produce	−	+	+
Hydrocracking	Very slow	Exo	No	++	++	Consume	+	−	+

3.5 CONCLUSION

Finally, recent studies have discovered that special reactor arrangements, particularly fluidized bed reactors, have the potential to reduce the generation of excessively cracked products and optimize catalytic reforming. However, for this direct scaling-up approach to work on an industrial scale, it is necessary to successfully overcome various obstacles. These obstacles include improving the phase of the fluidized structure, enhancing the raw materials mixture, guaranteeing an efficient heat transfer, improving product separation, and increasing the resistance to fouling. In addition to the operative aspects of the reactor system, other practical challenges involve the heat cost and the energy per unit produced, as well as the potentially high demand for technologies that facilitate hydrogen recovery. In particular, the relationship between the catalysts and the hydrothermal effects, especially in cases that involve high mass content and low hydrogen content, in addition to the use of membrane and liquid catalytic reactants, presents stability challenges for catalyst-dependent systems.

All of these are considered a priority area of the research about the naphtha catalytic reforming, including the secondary CC reactions. This is because they have the potential to produce branched isoparaffins with high octane ratings and achieve significant yields in advanced reforming processes. In the last decades there have been a lot of advances in the catalysis field, but there is still a lot to perfectionate and improve in this field, under the cost-benefit premise, and the recently added social and environmental impacts.

REFERENCES

1. Jones, D. S.; Pujadó, P. R.; *Handbook of petroleum processing*; Springer Science & Business Media: Países Bajos, 2006.
2. Maples, R.E.; *Petroleum refinery process economics*, 2nd ed.; Pennwell books: Tulsa, Okla. USA, 2000.
3. Rahimpour, M. R.; Jafari, M.; Iranshahi, D.; Progress in catalytic naphtha reforming process: A review. *Applied Energy*. 2013, 109, 79–93.
4. Demirbas, A.; Balubaid, M. A.; Basahel, A. M.; Ahmad, W.; Sheikh, M. H.; Octane Rating of Gasoline and Octane Booster Additives. *Petroleum Science and Technology*. 2015, 33:11, 1190–1197.
5. Rahimpour, M. R.; Jafari, M.; Iranshahi, D.; Progress in catalytic naphtha reforming process: A review. *Applied Energy*. 2013, 109, 79–93.
6. Liu, Y. A.; Chang, A. F.; Pashikanti, K.; *Petroleum Refinery Process Modeling: Integrated Optimization Tools and Applications*; Wiley-VCH: Weinheim, Germany, 2018.
7. Vargas, S. Z.; Olaya, J. A.; Elaboración de curvas de destilación de hidrocarburos líquidos pertenecientes a la sub-cuenca Neiva: método ASTM D86-04. *Ingeniería y Región*. 2012, 9, 37–45.
8. Ramadhan, O. M.; Al-Hyali, E. A.; New experimental and theoretical relation to determine the research octane number (RON) of authentic aromatic hydrocarbons could be present in the gasoline fraction. *Petroleum Science and Technology*. 1999, 17(5-6), 623–635.
9. Albahri, T. A.; Structural group contribution method for predicting the octane number of pure hydrocarbon liquids. *Industrial & Engineering Chemistry Research*. 2003, 42(3), 657–662.

10. Demirbas, A.; Balubaid, M. A.; Basahel, A. M.; Ahmad, W.; Sheikh, M. H.; Octane rating of gasoline and octane booster additives. *Petroleum Science and Technology*. 2015, 33(11), 1190–1197.

11. Balaban, A. T.; Kier, L. B.; Joshi, N.; Structure-property analysis of octane numbers for hydrocarbons (alkanes, cycloalkanes, alkenes). *MATCH Communications in Mathematical and in Computer Chemistry*. 1992, 28, 13–27.

12. Cerri, T.; D'Errico, G.; Onorati, A.; Experimental investigations on high octane number gasoline formulations for internal combustion engines. *Fuel*. 2013, 111, 305–315.

13. Antos, G. J.; Aitani, A. M.; Parera, J. M.; *Catalytic naphtha reforming: science and technology*; Marcel Dekker Inc: New York, USA, 1995.

14. Sotelo-Boyás, R.; Froment, G. F.; Fundamental kinetic modeling of catalytic reforming. *Industrial & Engineering Chemistry Research*. 2009, 48(3), 1107–1119.

15. Abeladim, A.; Etude de la Transformation desNaphtenes en Présence de Catalyseurs de Reformage Catalytique, Ph.D. Thesis. Ecole Nationale Supérieure du Pétrole et des Moteurs, Rueil Malmaison, 1982.

16. Alvarez-Herrera, C.; Etude Cinétique de la Déshydrocylisation du N-heptane sur Catalyseur Platine sur Alumine, Ph.D. Thesis. Poitiers University, Poitiers, 1977.

17. Díaz-Velázquez, H.; Likhanova, N.; Aljammal, N.; Verpoort, F.; Martínez-Palou, R.; New insights into the progress on the isobutane/butene alkylation reaction and related processes for high-quality fuel production. *A Critical Review. Energy & Fuels*. 2020, 34(12), 15525–15556.

18. Kim, K. J.; Kim, K. Y.; Rhim, G. B.; Youn, M. H.; Lee, Y. L.; Chun, D. H.; Roh, H. S.; Nano-catalysts for gas to liquids: A concise review. *Chemical Engineering Journal*. 2023, 143632.

19. Leung, S. L.; Exploiting Diffusional Constraints in Microporous Materials for Methane Reforming Reactions with Insights from Isotopic Exchange Experiments, Doctoral dissertation. University of California, Berkeley, 2021.

20. Monai, M.; Gambino, M.; Wannakao, S.; Weckhuysen, B. M.; Propane to olefins tandem catalysis: a selective route towards light olefins production. *Chemical Society Reviews*. 2021, 50(20), 11503–11529.

21. Martínez, J.; Zúñiga-Hinojosa, M. A.; Ruiz-Martínez, R. S.; A thermodynamic analysis of naphtha catalytic reforming reactions to produce high-octane gasoline. *Processes*. 2022, 10(2), 313.

22. Merezhko, N.; Tkachuk, V.; Romanchuk, V.; Rechun, O.; Zagoruiko, V.; High-Octane Fuel Compositions Based on Petroleum and Biocomponents. In *Grabchenko's International Conference on Advanced Manufacturing Processes*. Eds.; Springer International Publishing: Odessa, Ukraine, 2020; 685–694.

23. Petrova, D. A.; Gushchin, P. A.; Ivanov, E. V.; Lyubimenko, V. A.; Kolesnikov, I. M.; Modelling industrial catalytic reforming of low octane gasoline. *Chemistry and Technology of Fuels and Oils*. 2021, 57, 143–159.

24. Nwaobi, R. (2021). Oil Refinery Using Catalytic Reforming Unit, Mater dissertation Siberian Federal University, Krasnoyarsk, 2021.

25. Smolikov, M. D.; Kir'yanov, D. I.; Shkurenok, V. A.; Bikmetova, L. I.; Belopukhov, E. A.; Yablokova, S. S.; Kleimenov, A. V.; Integrated processes of the reforming and isomerization of gasoline fractions for the production of environmentally friendly motor gasolines. *Catalysis in Industry*. 2022, 14(3), 268–282.

26. Abdellatief, T. M.; Ershov, M. A.; Kapustin, V. M.; Abdelkareem, M. A.; Kamil, M.; Olabi, A. G.; Recent trends for introducing promising fuel components to enhance the anti-knock quality of gasoline: A systematic review. *Fuel*. 2021, 291, 120112.

27. Samad, A.; Ahmad, I.; Kano, M.; Caliskan, H.; Prediction and optimization of exergetic efficiency of reactive units of a petroleum refinery under uncertainty through artificial neural network-based surrogate modeling. *Process Safety and Environmental Protection*. 2023, 177, 1403–1414.

28. Kapner, R. S.; Lannus, A.; Thermodynamic analysis of energy efficiency in catalytic reforming. *Energy*. 1980, 5(8-9), 915–924.

29. Gomes, S. R.; Bion, N.; Blanchard, G.; Rousseau, S.; Bellière-Baca, V.; Harlé, V.; Epron, F.; Thermodynamic and experimental studies of catalytic reforming of exhaust gas recirculation in gasoline engines. *Applied Catalysis B: Environmental*. 2011, 102(1-2), 44–53.

30. Rahimpour, M. R.; Jafari, M.; Iranshahi, D.; Progress in catalytic naphtha reforming process: A review. *Applied Energy*. 2013, 109, 79–93.

31. Yusuf, A. Z.; John, Y. M.; Aderemi, B. O.; Patel, R.; Mujtaba, I. M.; Effect of hydrogen partial pressure on catalytic reforming process of naphtha. *Computers & Chemical Engineering*. 2020, 143, 107090.

32. Shakor, Z. M.; Abdul-Razak, A. A.; Sukkar, K. A.; A detailed reaction kinetic model of heavy naphtha reforming. *Arabian Journal for Science and Engineering*. 2020, 45(9), 7361–7370.

33. Meng, S.; Li, W.; Xu, H.; Li, Z.; Li, Y.; Jarvis, J.; Song, H.; Non-thermal plasma assisted catalytic reforming of naphtha and its model compounds with methane at near ambient conditions. *Applied Catalysis B: Environmental*. 2021, 297, 120459.

34. Soltanali, S.; Mohaddecy, S. R.; Mashayekhi, M.; Rashidzadeh, M.; Catalytic upgrading of heavy naphtha to gasoline: Simultaneous operation of reforming and desulfurization in the absence of hydrogen. *Journal of Environmental Chemical Engineering*. 2020, 8(6), 104548.

35. Zhang, P.; Yang, Y.; Li, Z.; Liu, B.; Hu, C.; Preparation, characterization and naphtha aromatization performance of the catalytic reforming catalyst Pt/MY (M= Mg, Ba or Ce). *Catalysis Today*. 2020, 353, 146–152.

36. Ren, J.; Cao, J. P.; Yang, F. L.; Liu, Y. L.; Tang, W.; Zhao, X. Y.; Understandings of catalyst deactivation and regeneration during biomass tar reforming: A crucial review. *ACS Sustainable Chemistry & Engineering*. 2021, 9(51), 17186–17206.

37. Greluk, M.; Rotko, M.; Turczyniak-Surdacka, S.; Comparison of catalytic performance and coking resistant behaviors of cobalt-and nickel based catalyst with different Co/Ce and Ni/Ce molar ratio under SRE conditions. *Applied Catalysis A: General*. 2020, 590, 117334.

38. Alcala, R.; Dean, D. P.; Chavan, I.; Chang, C. W.; Burnside, B.; Pham, H. N.; Datye, A. K.; Strategies for regeneration of Pt-alloy catalysts supported on silica for propane dehydrogenation. *Applied Catalysis A: General*. 2023, 658, 119157.

39. Choudhury, D.; Das, R.; Tripathi, A. K.; Priyadarshani, D.; Neergat, M.; Kinetics of hydrogen evolution reactions in acidic media on Pt, Pd, and MoS2. *Langmuir*. 2022, 38(14), 4341–4350.

40. Raju, S.; Volume effects of alloying: a thermodynamic perspective. *Transactions of the Indian Institute of Metals*. 2022, 75(4), 1031–1041.

41. Said-Aizpuru, O.; Allain, F.; Diehl, F.; Farrusseng, D.; Joly, J. F.; Dandeu, A.; A naphtha reforming process development methodology based on the identification of catalytic reactivity descriptors. *New Journal of Chemistry*. 2020, 44(18), 7243–7260.

42. Said-Aizpuru, O.; Allain, F.; Dandeu, A.; Diehl, F.; Farrusseng, D.; Joly, J. F.; Kinetic modelling of Pt/γ-Al 2 O 3–Cl catalysts formulation changes in n-heptane reforming. *Reaction Chemistry & Engineering*. 2021, 6(6), 1079–1091.

43. Ali, S. A.; Alshareef, A. H.; Theravalappil, R.; Alasiri, H. S.; Hossain, M. M.; Molecular kinetic modeling of catalytic naphtha reforming: a review of complexities and solutions. *Catalysis Reviews.* 2023, 65(4), 1358–1411.

44. Chehade, A. M.; Daher, E. A.; Assaf, J. C.; Riachi, B.; Hamd, W.; Simulation and optimization of hydrogen production by steam reforming of natural gas for refining and petrochemical demands in Lebanon. *International Journal of Hydrogen Energy.* 2020, 45(58), 33235–33247.

45. Zagoruiko, A. N.; Belyi, A. S.; Smolikov, M. D.; Thermodynamically Consistent Kinetic Model for the Naphtha Reforming Process. *Industrial & Engineering Chemistry Research.* 2021, 60(18), 6627–6638.

46. Mokheimer, E. M.; Shakeel, M. R.; Harale, A.; Paglieri, S.; Mansour, R. B.; Fuel reforming processes for hydrogen production. *Fuel.* 2024, 359, 130427.

47. Fowles, M.; Carlsson, M.; Steam reforming of hydrocarbons for synthesis gas production. *Topics in Catalysis.* 2021, 64(17), 856–875.

48. Yoo, J. Y.; Lee, J.; Han, G.; Harale, A.; Katikaneni, S.; Paglieri, S. N.; Bae, J.: On-site hydrogen production using heavy naphta by maximizing the hydrogen output of a membrane reactor system. *Journal of Power Sources.* 2021, 508, 230332.

49. Zhou, Y.; Huang, Z.; Zhang, K.; Yang, M.; Zhan, J.; Liu, M.; Zhou, Y.; Economic analysis of hydrogen production and refueling station via molten-medium-catalyzed pyrolysis of natural gas process. *International Journal of Hydrogen Energy.* 2024, 62, 1205–1213.

50. Muhammed, N. S.; Gbadamosi, A. O.; Epelle, E. I.; Abdulrasheed, A. A.; Haq, B.; Patil, S.; Kamal, M. S.; Hydrogen production, transportation, utilization, and storage: Recent advances towards sustainable energy. *Journal of Energy Storage.* 2023, 73, 109207.

51. Mehanovic, D.; Peloquin, J. F.; Dufault, J. F.; Fréchette, L.; Picard, M.; Comparative techno-economic study of typically combustion-less hydrogen production alternatives. *International Journal of Hydrogen Energy.* 2023, 48(22), 7945–7958.

52. Weber, J. M.; Lapkin, A. A.; Micro-kinetics analysis based on partial reaction networks to compare catalysts performances for methane dry reforming reaction. *Chemical Engineering Journal.* 2023, 466, 143212.

53. Velisoju, V. K.; Kulkarni, S. R.; Cui, M.; Rabee, A. I.; Paalanen, P.; Rabeah, J.; Castaño, P.; Multi-technique operando methods and instruments for simultaneous assessment of thermal catalysis structure, performance, dynamics, and kinetics. *Chem Catalysis.* 2023.

54. Kim, K. J.; Kim, K. Y.; Rhim, G. B.; Youn, M. H.; Lee, Y. L.; Chun, D. H.; Roh, H. S.; Nano-catalysts for gas to liquids: A concise review. *Chemical Engineering Journal.* 2023, 143632.

55. Eyal, A.; Tartakovsky, L.; Second-law analysis of the reforming-controlled compression ignition. *Applied Energy.* 2020, 263, 114622.

56. Motagamwala, A. H.; Dumesic, J. A.; Microkinetic modeling: a tool for rational catalyst design. *Chemical Reviews.* 2020, 121(2), 1049–1076.

57. Hu, Q.; Pang, S.; Wang, D.; In-depth insights into mathematical characteristics, selection criteria and common mistakes of adsorption kinetic models: A critical review. *Separation & Purification Reviews.* 2022, 51(3), 281–299.

58. Teimouri, Z.; Abatzoglou, N.; Dalai, A. K.; Kinetics and selectivity study of Fischer–Tropsch synthesis to C5+ hydrocarbons: a review. *Catalysts.* 2021, 11(3), 330.

59. Ancheyta-Juárez, J.; López-Isunza, F.; Analysis of deactivation models based on Time-on-Stream (TOS) Theory for fluid catalytic cracking process. *Revista de la Sociedad Química de México.* 2000, 44(3), 183–187.

60. Beltramini, J. N.; Cabrol, R. A.; Churin, E. J.; Figoli, N. S.; Martinelli, E. E.; Parera, J. M.; Catalyst deactivation by naphthas doped with hydrocarbons. *Applied Catalysis.* 1985, 17(1), 65–74.

61. Zaynullin, R. Z.; Koledina, K. F.; Gubaydullin, I. M.; Akhmetov, A. F.; Koledin, S. N.; Kinetic model of catalytic gasoline reforming with consideration for changes in the reaction volume and thermodynamic parameters. *Kinetics and Catalysis.* 2020, 61, 613–622.

62. Azizan, M. T.; Aqsha, A.; Ameen, M.; Syuhada, A.; Klaus, H.; Abidin, S. Z.; Sher, F.; Catalytic reforming of oxygenated hydrocarbons for the hydrogen production: an outlook. *Biomass Conversion and Biorefinery.* 2020, 1–24.

63. Amiri, T. Y.; Ghasemzageh, K.; Iulianelli, A.; Membrane reactors for sustainable hydrogen production through steam reforming of hydrocarbons: A review. *Chemical Engineering and Processing-Process Intensification.* 2020, 157, 108148.

64. Guo, Q.; Geng, J.; Pan, J.; Zou, L.; Tian, Y.; Chi, B.; Pu, J.; Brief review of hydrocarbon-reforming catalysts map for hydrogen production. *Energy Reviews.* 2023, 100037.

65. Babaqi, B. S.; Takriff, M. S.; Othman, N. T. A.; Kamarudin, S. K.; Yield and energy optimization of the continuous catalytic regeneration reforming process based particle swarm optimization. *Energy.* 2020, 206, 118098.

66. Xiao, Z.; Zhang, C.; Huang, S.; Zhang, S.; Tan, X.; Lian, Z.; Wang, D.; A comprehensive review on steam reforming of liquid hydrocarbon fuels: Research advances and Prospects. *Fuel.* 2024, 368, 131596.

67. Iranshahi, D.; Rahimpour, M. R.; Asgari, A.; A novel dynamic radial-flow, spherical-bed reactor concept for naphtha reforming in the presence of catalyst deactivation. *International Journal of Hydrogen Energy.* 2010, 35(12), 6261–6275.

68. Rahimpour, M. R.; Iranshahi, D.; Pourazadi, E.; Paymooni, K.; Bahmanpour, A. M.; The aromatic enhancement in the axial-flow spherical packed-bed membrane naphtha reformers in the presence of catalyst deactivation. *AIChE Journal.* 2011, 57(11), 3182–3198.

69. Demirbas, A.; Competitive liquid biofuels from biomass. *Applied Energy.* 2011, 88(1), 17–28.

70. Roddy, D. J.; Development of a CO_2 network for industrial emissions. *Applied Energy.* 2012, 91(1), 459–465.

71. Ancheyta-Juárez, J.; Villafuerte-Macías, E.; Kinetic modeling of naphtha catalytic reforming reactions. *Energy & Fuels.* 2000, 14(5), 1032–1037.

72. Ciapetta, F. G.; Wallace, D. N.; Catalytic naphtha reforming. *Catalysis Reviews.* 1972, 5(1), 67–158.

73. Zakari, A. Y.; Aderemi, B. O.; Patel, R.; Mujtaba, I. M.; Study of industrial naphtha catalytic reforming reactions via modelling and simulation. *Processes.* 2019, 7(192), 1–26.

74. Pieck, C. L.; Sad, M. R.; Parera, J. M.; Chlorination of Pt–Re/Al_2O_3 during naphtha reforming. *Journal of Chemical Technology & Biotechnology: International Research in Process, Environmental AND Clean Technology.* 1996, 67(1), 61–66.

75. Mariscal, R.; Fierro, J. L.; Yori, J. C.; Parera, J. M.; Grau, J. M.; Evolution of the properties of PtGe/Al_2O_3 reforming catalysts with Ge content. *Applied Catalysis A: General.* 2007, 327(2), 123–131.

76. Wei, W.; Bennett, C. A.; Tanaka, R.; Hou, G.; Klein, M. T.; Detailed kinetic models for catalytic reforming. *Fuel Processing Technology.* 2008, 89(4), 344–349.

77. Sotelo-Boyás, R.; Froment, G. F.; Fundamental kinetic modeling of catalytic reforming. *Industrial & Engineering Chemistry Research.* 2009, 48(3), 1107–1119.

78. Zainullin, R. Z.; Koledina, K. F.; Akhmetov, A. F.; Gubaidullin, I. M.; Kinetics of the catalytic reforming of gasoline. *Kinetics and Catalysis.* 2017, 58, 279–289.

79. Rahimpour, M. R.; Jafari, M.; Iranshahi, D.; Progress in catalytic naphtha reforming process: A review. *Applied Energy.* 2013, 109, 79–93.

4 Isomerization/ Hydroisomerization

Araceli Vega Paz
Mexican Petroleum Institute, Mexico City, México

4.1 INTRODUCTION

Isomerization is very important in the petroleum industry since it transforms paraffins (linear alkanes) into isoparaffins[1] to produce fuels of high commercial value. One of the main fuels produced is gasoline, and it is desirable to achieve a high-octane rating.

Isomerization is a chemical process where one molecule is transformed into another with the same number of carbon atoms and molecular formula, but a different structural formula or three-dimensional structure (Figure 4.1).

The general description of isomerization indicates the breaking of double bonds of the olefinic type using an acid catalyst, generating a carbocation that is rearranged to obtain the branches in the linear structure;[2] whereas in hydroisomerization, the catalyst used is of the metallic type (for dehydrogenation), which extracts hydrogen from the saturated hydrocarbon and forms double bonds.

It was in 1930 when the first catalysts for the isomerization of paraffin to obtain isobutane, UOP technology, were developed using aluminum chloride. The first butane isomerization plant was developed in 1941, but it was expensive to maintain; later hydrogen fluoride (HF) was implemented as a liquid acid catalyst in the production of higher-octane gasoline, starting from isobutane with isobutylene to generate isooctane. In the 1950s–1960s, the need for highly efficient fuels increased, and to achieve this, the UOP-PENEX process was applied to isomerize light naphtha, achieving an increase

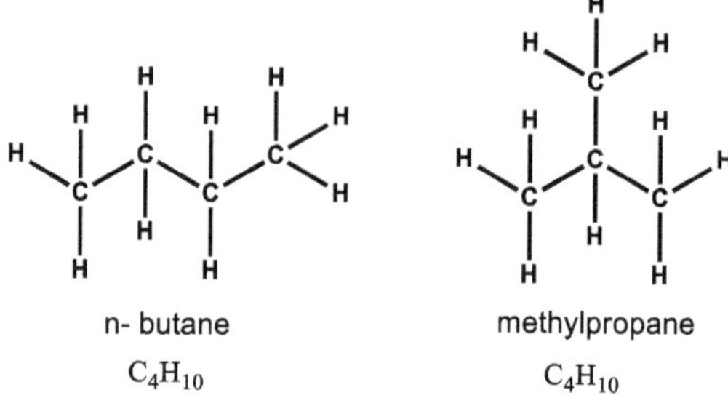

n- butane

C_4H_{10}

methylpropane

C_4H_{10}

FIGURE 4.1 Chemical structure of n-butane and its structural isomer.

DOI: 10.1201/9781003517283-4

in octane number between 12 and 20. Since 1960, an advancement has been identified in the catalysts used for isomerization of light naphtha, with technologies such as Akzo Nobel NV (Albemarle), Axens, NPP Neftekhim, Süd-Chemie AG (Clariant), Shell Chemicals LP, and UOP (Honeywell UOP LLC).

The production of isoalkanes can occur in three ways:

1. Isomerization of n-pentane and n-hexane from crude distillation (light direct distillation, LSR) to the corresponding isoalkanes (research octane number (RON) ca. 80).
2. Dimerization plus hydrogenation (indirect alkylation), which converts isobutene to trimethylpentanes (RON 100). When the feed is a butene mixture, the octane ratings are lower (RON ca. 80) due to the production of dimethylhexanes.
3. Alkylation of C3–C5 alkenes with isobutane to give C5–C12 isoalkanes (RON 92–96). Of these three, isomerization and alkylation are key processes for improving gasoline feedstock.

4.2 MECHANISM REACTION

4.2.1 ISOMERIZATION REACTION

To initiate the isomerization reaction, it requires an acidic catalyst that is protonated at an acidic center allowing the formation of a carbocation (carbenium) at the vinyl bond (-C=C-) of a molecule. This can occur by rearranging the carbon backbone of the carbocation to an unsaturated two-branched molecule and then hydride transferring between n-alkane molecules and the isomerized carbenium ions to produce the isoalkane and new carbenium ions.[2]

4.2.2 HYDROISOMERIZATION REACTION

This occurs when a linear alkane forms the corresponding alkene by dehydrogenation over the metal salt.

The vinyl bond of the alkene undergoes protonation at the acidic center to form the corresponding carbenium ion, which can lead to reactions such as isomerization and can occur in two ways: Monofunctional catalysis (pure acid) or bifunctional catalysis, both of which involve carbenium ion formation.

4.2.3 MONOFUNCTIONAL CATALYSIS

Also known as acid catalysis, it is characterized by high selectivity toward the corresponding isomers and a high energy barrier.

The reaction mechanism of an alkane includes three main steps (Figure 4.2), with the first being the formation of the carbenium ion: The alkane undergoes deprotonation upon interaction with the Brönsted acid site of the catalyst, forming the carbenium ion.

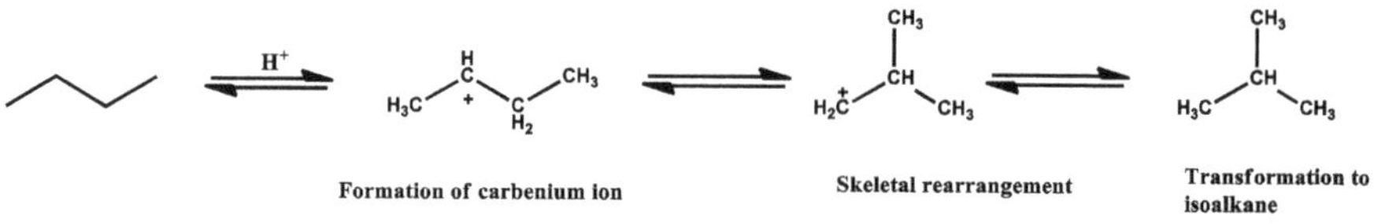

Formation of carbenium ion   **Skeletal rearrangement**   **Transformation to isoalkane**

FIGURE 4.2   Main steps in the isomerization of an alkane.

Dehydrogenation

$$\text{n-}C_5H_{12} \rightleftharpoons \text{n-}C_5H_{10} + H_2$$

Skeletal rearrangement   **Hydrogenation**

$$\text{n-}C_5H_{10} + H^+ \rightleftharpoons \text{n-}C_5H_{11}^+ \longrightarrow \text{Iso-}C_5H_{11}^+ \rightleftharpoons \text{Iso-}C_5H_{10} + H^+ \rightleftharpoons \text{Iso-}C_5H_{12}$$

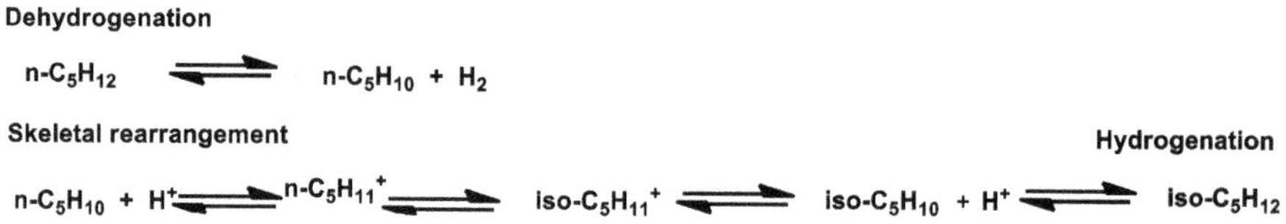

FIGURE 4.3   Main steps in the isomerization of an alkane with a bifunctional catalyst.

In the second step (skeleton rearrangement), the carbene ion undergoes a transposition due to the formation of an unstable intermediate (protonated cyclopropane).

The third step is isoalkane formation: The iso-R carbene in contact with an n-R molecule favors isoalkane formation by deprotonation of the alkane.[3]

4.2.4 BIFUNCTIONAL CATALYSIS

The catalyst has Lewis and Brönsted acid sites. The Lewis sites favor the dehydrogenation reaction of n-alkanes, forming the corresponding alkene, and the Lewis sites have the property of inhibiting the formation of coke. The alkene formed favors the formation of the corresponding carbene ion due to the Brönsted acid sites, followed by the structural rearrangement step and formation of the isoalkane, Figure 4.3.

A high concentration of alkenes favors the nonselective bifunctional pathway, and the content of feed olefins and/or the activation rate of the n-alkane affects the isomerization pathway.[3]

4.3 IMPORTANCE OF ISOMERIZATION AS A REFINING PROCESS

Isomerization of light naphtha (typically C5 and C6) as a refining process in the petroleum industry is of great importance in the production of petroleum derivatives (Figure 4.4) because it is a simple and very profitable technology to increase the octane number, achieving an increase of 10 to 20 octanes, increasing the octane number, and reducing the benzene in the gasoline pool. Of course, it is a function of the gas feed.

As mentioned above, it is of particular interest to produce higher-quality gasoline, with high octane rating and reduced emissions of pollutants into the environment. The isomerization of naphtha (low quality) in the production of high-quality feedstocks is of economic importance.

It should be noted that the isomerization process is complex and requires constant monitoring and good control to ensure a smooth and efficient reaction to improve yields and reduce costs.

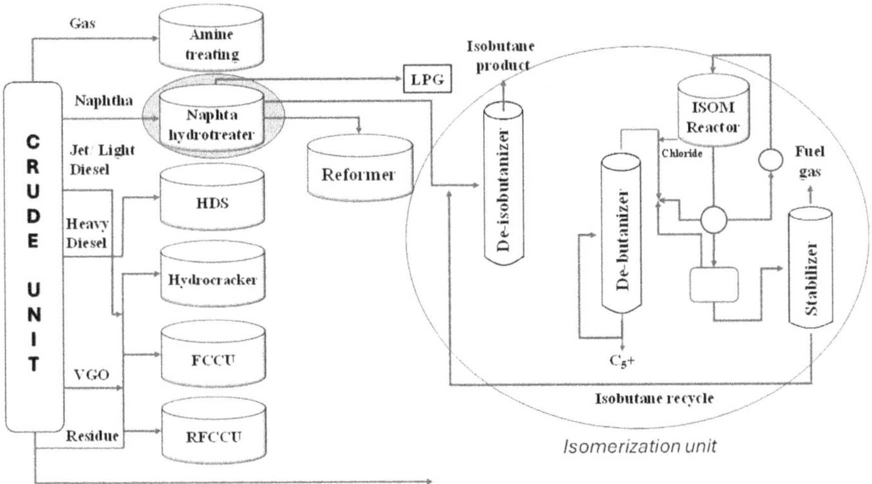

FIGURE 4.4 Refining diagram showing the location of the isomerization process.

4.4 CATALYSTS FOR ISOMERIZATION

The isomerization reaction is a complex, exothermic process with a fast reaction rate, requiring a suitable catalyst to promote isomerization and minimize side reactions under mild conditions.

The isomerization reaction takes place in a high-temperature, high-pressure reactor packed with the acid catalyst, of which there are several types.

4.4.1 SUMMARY OF COMPOSITION, APPLICATION TEMPERATURE, ADVANTAGES, AND DISADVANTAGES OF THE DIVERSE TYPES OF CATALYSTS

Catalysts for the hydroisomerization reaction of light n-alkanes have evolved and there are several generations of them.

The first generation corresponds to Friedel–Crafts catalyst ($AlCl_3$), with catalytic activity between 80°C and 100°C and high sensitivity to impurities and corrosives.

The second generation is formed by Pt supported on functional alumina at temperatures between 350°C and 500°C, favoring cracking and coke deposition and generating high amounts of C1 to C4 gases.

Third-generation catalysts correspond to platinum supported on chlorinated alumina and have high acidity; therefore, the operating temperature is between 80°C and 100°C.

Fourth-generation catalysts correspond to bifunctional catalysts with an acid function provided by zeolite and a dehydrogenation–hydrogenation function provided by platinum, with catalytic activity at 250°C and more resistance to impurities.

Fifth-generation catalysts consist of zirconia: Sulfonated and tungstenated.

Table 4.1 shows the catalysts according to their classification.

TABLE 4.1

Types of Hydroisomerization Catalysts

Catalyst	Composition	Temperature, °C	Advantage	Disadvantage
First generation	Friedel–Crafts AlCl$_3$	80–100	Hight activity	Highly sensitive to moisture, non-reusable
Second generation	Metal/support	350–500	Easy application	Promotes cracking and rapid deactivation
Third generation	Metal/support halogenated	120–160	Enhanced acidity by halogenation of supported alumina	Chloride addition is necessary for catalytic activity. Sensitive to poisons
Fourth generation	Bifunctional, zeolite-based	250–340	Easy to use and no pretreatment of raw material required	Requires high temperatures and needs high H$_2$/ hydrocarbon ratios
Fifth generation	Bifunctional, zirconia based	50–170	High activity and water tolerance.	Requires higher H$_2$/ hydrocarbon ratio

Chlorinated Aluminas

Alumina is aluminum oxide (Al$_2$O$_3$) (Figure 4.5). It is one of the most important components in the composition of clays and glazes, making them resistant. It is the common form of crystalline aluminum oxide (corundum), and it is the most thermodynamically stable form. Ionic oxygen is part of the compact hexagonal structure where aluminum cations fill two-thirds of the octahedral interstices; therefore, Al^{3+} is octahedral. Aluminum oxide is found in various phases such as γ and η, monoclinic, hexagonal, orthorhombic, etc (maybe tetragonal or orthorhombic).[4]

The catalytic properties of alumina include surface structure, associated with its electronic and proton distribution, specific area, and pore distribution. The Brönsted and Lewis acidity sites of aluminas (porous material) are enhanced by chlorine and are used in the hydroisomerization of light naphtha, The incorporation of platinum (Pt/Al$_2$O$_3$) favors their activity and selectivity since it reduces the deposition of coke.

Chlorinated aluminas work as catalysts at temperatures in the range of 130–170°C, achieving RONs ranging from 88 to 92. Among their disadvantages is the sensitivity

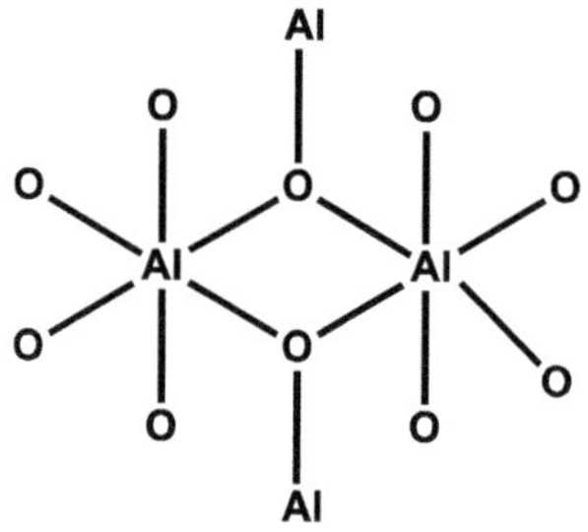

FIGURE 4.5   General chemical structure of alumina.

to water, nitrogen, and sulfur; in addition, during isomerization, to maintain the alumina with catalytic activity, chlorides must be added; caustic purification processes are also required. The useful life of these materials is 2–3 years, without the possibility of regeneration.

Since these catalysts are very sensitive to feed impurities such as sulfur, nitrogen, and water, removal of these impurities by specific purification steps is required, which usually increases the cost of the unit. This process requires the addition of chlorides to maintain catalyst activity and the removal of hydrogen chloride in a caustic scrubber, which requires the disposal of spent caustic and the associated disposal problem for refineries.

Zeolites

Zeolites are aluminosilicates (Figure 4.6), three-dimensional structures, consisting of aluminum, silicon, and oxygen, forming tetrahedrons interconnected by shared oxygens, including K, Na, Ca, Li, Mg, Sr, and Ba and n-water molecules (Figure 4.7). Zeolites can be represented by the general formula $M_nAl_xSi_yO_2(x+y).wH_2O$ (M=exchangeable cation of Group I or II with valence n, x = number of Al, y = number of Si, and w=H_2O molecules).

Zeolites have channels and cavities of nano and subnanometer dimensions; that is, they are nanoporous structures, which means high surface area.

The cavities and channels have water molecules by inclusion. When these molecules are removed, the channels are dehydrated, and in the presence of another type of chemical structure with the right diameter are easily adsorbed by the channels and cavities (molecular sieve). Zeolites undergo isomorphic substitution of silicon atoms

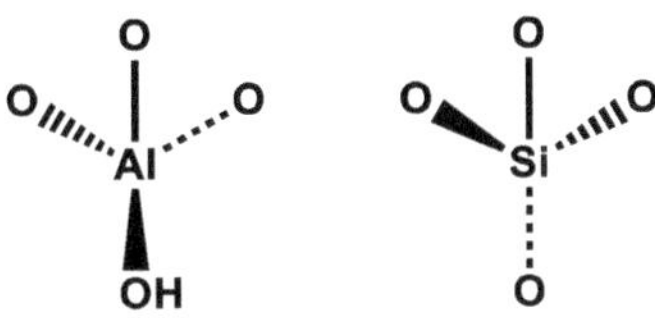

FIGURE 4.6   General chemical structure of zeolites.

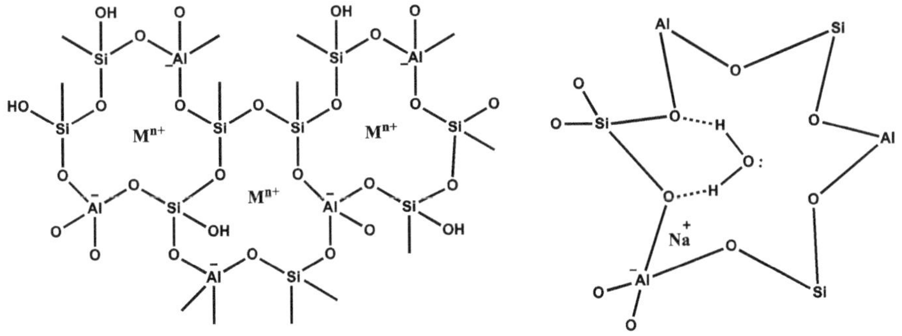

FIGURE 4.7   Structure of a zeolite, representation of the inclusion.

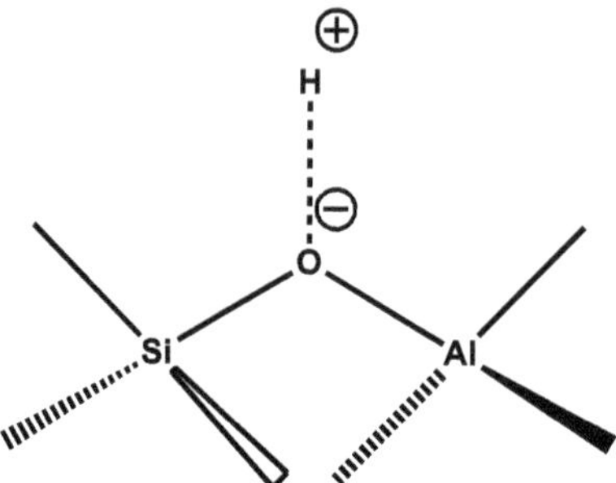

FIGURE 4.8 Representation of Brönsted acid sites in zeolites.

by tetravalent aluminum generating a negative charge; such an exchange is favored in the microporous crystalline structure; in other words, the Si^{4+}/Al^{3+} ratio of the structure of each zeolite determines the charge of the network, the greater the amount of Si^{4+} cations, the greater the negative charge favoring ion exchange. However, characteristics such as micropore size, channel dimension, SiO_2/Al_2O_3 ratio, and connectivity determine the catalytic efficiency.

The acidity of these materials is a function of the negative charge associated with the aluminum atoms when the electronegativity is compensated by the protons of the oxygens shared between silicon and aluminum (Figure 4.8).

The H^+ cation is labile; therefore, it is considered a Brönsted acid site. It is this cation that can be easily released.

Within the zeolite structure, there are also Lewis acid sites, which are formed by the loss of water molecules (Figure 4.9).

Zeolites have the disadvantage of aluminum expulsion at high temperatures which is generated in the form of oxides and hydroxides.[5]

Zeolites can be classified by pore size:[6]

- Small 3.5–4.5 Å, Zeolite A: Small molecules such as those in gasoline.
- Medium 4.5–6.0 Å, Zeolite Synthetic Mobil (ZSM) 5: They have selectivity to heavy n-paraffins.
- Large 6.0–8.0 Å, Zeolite X and Y: For light and heavy paraffins.

Among medium pore zeolites, one of the most studied is ZSM 5 of the pentasil family, constituted by two SiO_4 or AlO_4^- units and linked by shared oxygens. The high Si/Al ratio confers thermal stability, hydrophobicity, and low ion exchange. Besides, the acid strength is high and can be modified for aluminum extraction.[7]

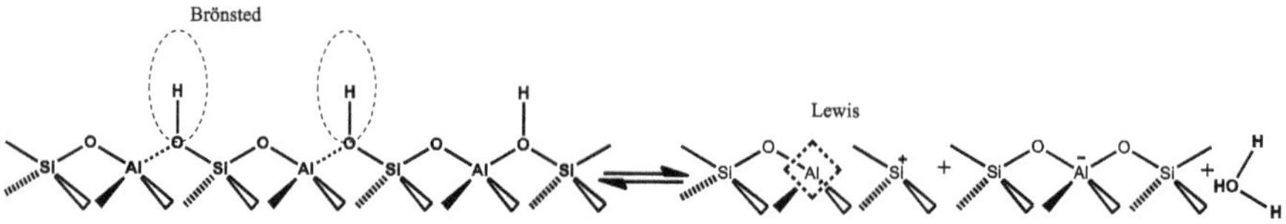

FIGURE 4.9 Degradation of a zeolite structure.

It has a high internal surface, good thermal stability, high adsorption and ionic exchange capacity, and higher resistance to deactivation than other types of zeolites because structurally its cavities are small and the channels are interconnected in three directions, thus avoiding blocking of a pore initiating the deactivation of all the acid centers.[8]

Pt/zeolite catalysts have lower catalytic activity than Pt/Al_2O_3-Cl and Pt/ZrO_2-SO_4 type catalysts in the hydroisomerization reaction of light naphtha. To improve the catalytic efficiency of Pt/zeolites, the reaction temperature is increased with a decrease in octane number due to thermodynamic equilibrium.

The advantage of these materials is their high tolerance to water and low sulfur concentrations present in the reactor feed; besides, it is possible to recycle them several times.

The acidity gives the possibility to generate the carbene ion and to carry out the isomerization reaction by activation of the C–C and C–H bonds. So, they are used in the refining and petrochemical industry. To achieve a hydrogenating–dehydrogenating function, to dehydrogenate alkanes and hydrogenate alkenes, bifunctional catalysts are developed. Among the chemical elements of the periodic table that favor hydrogenation, those belonging to the d-block (transition metals) with partially filled d or f orbitals are identified. The most used are Pt, Pd, and Ni (group 10). Also used for this purpose are oxides and sulfides of Mo, W, Fe, and Co.[9] While Pd has shown its functionality in the hydroisomerization of n-butane,[10,11] Pt has high activity and selectivity to the hydroisomerization of n-alkanes with higher carbon number (C7, C8).

Hydroisomerization of n-hexane has been widely studied using Pt/zeolites.[12–17] Another type of zeolites used is the SAPO type (silicoaluminophosphates), which are synthesized from pseudo-boehmite (70% by weight of Al_2O_3), phosphoric acid (85% by weight of H_3PO_4), silica sol (30% by weight of SiO_2), and 0.3% of Na_2O.

The impregnation of SAPO 11 with Pd becomes a suitable material for the isomerization of large alkanes, such as n-tetradecane.[16] When SAPO 11 and SAPO 5 are enriched with bimetallic addition, mainly Pt-Ni and Pt-Pd, they favor the hydroisomerization reaction of n-hexane.[18]

The aim is to have more competitive zeolites from the catalytic point of view in the isomerization reaction, that is good control of the reaction conditions, adequate size of the metal species, and the required distribution to reach the largest number of acid sites with better access, which allows obtaining zeolites of nanometric dimensions (<100 nm), such as the mesoporous SBA-15 and MCM-41 zeolites with low acidity (due to the surface Si-OH groups), which when functionalized with Al improve the acidity and are suitable for obtaining n-octane by hydroisomerization.[19]

Large-pore zeolites include Y zeolites with faujasite (FAU) structure and micropores interconnected in all three directions. The pores are delimited by 12 tetrahedrally coordinated Si or Al atoms and have a diameter of about 7.4 mm, strong Brönsted acidity, and high hydrothermal stability. These materials are characterized by favoring hydrocracking.[20]

However, platinum impregnation results in Pt/zeolite Y, which can catalyze the hydroisomerization reaction of long-chain alkanes.[17]

4.5 METAL OXIDES

4.5.1 ZIRCONIUM OXIDE (ZrO_2)

Zirconia is a ceramic with surface acidity and basicity, oxide-reducing properties, porosity, surface area, thermal stability, high melting point, mechanical strength, low thermal conductivity, and corrosion resistance. Its acidity can be modified to provide a strong acid catalyst with efficiency in alkene isomerization reactions, hydrocracking, and alkylation.[21]

The synthesis of zirconia (ZrO_2) is achieved from a hydroxyl that is calcined to obtain the oxide, a crystalline material of the tetragonal or monoclinic type (depending on the heat treatment conditions), and existing studies indicate that the tetragonal crystalline form is the most active.[22]

One way to increase the acidity of metal oxides is by sulfonation, which produces superacids capable of catalyzing reactions at low temperatures.

Efforts continue to be directed towards having catalysts that meet the needs of the market and the increasingly demanding environmental regulations. In this sense there is an interest in anion-modified metal oxides with high acidity, in particular sulfonated zirconia, developed in 1962[23] with the ability to catalyze reactions at room temperature.[24–26]

4.5.2 SULFONATED ZIRCONIA (SO_4^{-2}-ZrO_2)

The suitability of the presence of Lewis and Brönsted sites in sulfonated zirconia for the isomerization of alkanes has been studied (Figure 4.10), where the factors involved in the catalytic activity are acidity, sulfur, activation temperature, water content, and deactivation, which occurs rapidly due to the modification of sulfate groups.[27–31] Superacidity favors high catalytic activity and can promote the isomerization reaction with metallic elements such as Co, Ni, Zn, Ag, Ga, and La, with the transition metals (block d) improving the conversion and selectivity. Although the effect of the promoter on catalytic activity has not been demonstrated, it is suggested that the distortion of the crystal lattice favors an increase in the mobility of ionic oxygen, creating vacancies and increasing electron diffusion; the isomerization reaction promoters work by dehydrogenation through oxy species, forming a greater number of olefinic species from alkanes.

The use of Fe and Mn does not affect the acidity of zirconia, but they do modify the oxide-reducing ability of the system.[32]

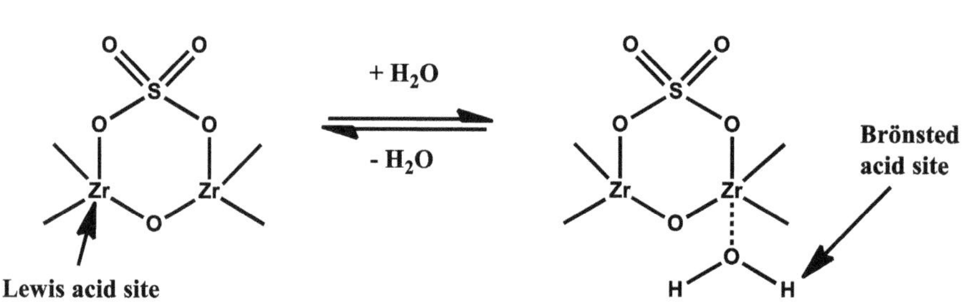

FIGURE 4.10 Lewis and Brönsted acid sites in sulfonated zirconia.

4.5.3 Reaction Mechanism of Sulfonated Zirconia

The extensive study of sulfonated zirconia is supported by several experimental works of catalytic evaluation, where the high activity of the catalyst allows for obtaining butene in situ by oxidative dehydrogenation of n-butane (Figure 4.11).[33,34]

When Pt is added to sulfonated zirconia, the material is more stable in the presence of hydrogen. Being a classical bifunctional catalyst, the isomerization of n-alkane starts with the dehydrogenation of the -C-C- bond due to the dissociative adsorption of a molecule on a metallic site, forming the corresponding olefin (C=C), which is protonated on the acid site forming the carbene ion, which by isomerization leads to the branched carbene ion, where the last step is the hydrogenation of the carbocation.[35]

4.5.4 Tungsten Zirconia (WO_3-ZrO_2)

To have super acid catalysts to promote the isomerization reaction of n-alkanes, free of sulfate to avoid the release of sulfur derivatives, tungsten zirconia (WO_3-ZrO_2) was developed from the hydrolysis of $ZrOCl.8H_2O$. The catalytic evaluation was carried out in a fixed-bed microcatalytic pulse reactor coupled with gas chromatography. The results of the evaluation when C5 is fed at 800°C indicate that the isopentane isomer (13% by weight) is being obtained; when the feed is a mixture of C4 and C5 at 100°C, the main product is isobutane (20% by weight).[36]

The standard for the use of WO_3-ZrO_2 with various modifications ranging from the synthesis method, promoters, support modifications, theoretical calculations, and new applications for achieving significant improvements in activity, selectivity, and stability are shown in Figure 4.12.[37–42]

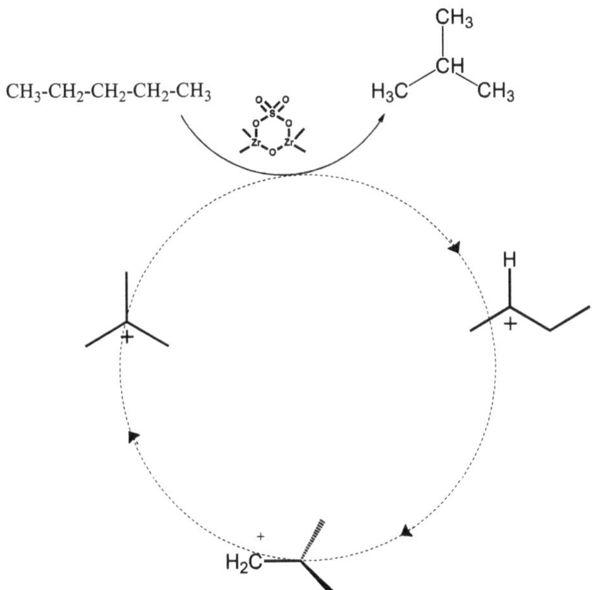

FIGURE 4.11 Dehydrogenation of butane, with the formation of carbenium ion when the catalyst is SO_4^{-2}-ZrO_2.

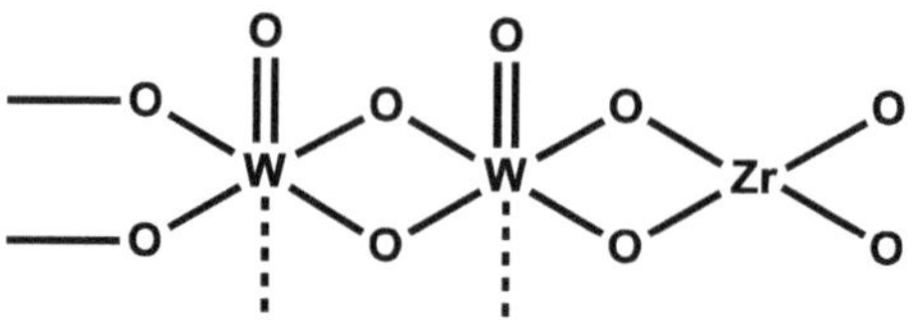

FIGURE 4.12   Representation of the tungsten zirconia structure.

The advantages of using WO_3-ZrO_2 in paraffin isomerization reactions include high thermal stability, easy regeneration, and low deactivation rates.[43–45]

The addition of noble metals to the WO_3-ZrO_2 structure to act as catalysts for the n-alkane hydroisomerization reaction favors stability and increases the catalytic activity in reactions where acidity is indispensable.[46]

The catalytic activity of zirconia is a function of the structure and size of the WO_x domain, which is necessary to maintain charge equilibrium and stabilize the carbocation intermediates.[47–49] While the Brønsted acid sites have a relationship with the WO_x groups, acidity enhancement (increase) is obtained by promoters such as aluminum[50–52] and other metals such as palladium which can hydrogenate.[53]

It has also been found that the addition of Pt, and Pd, increases the acidity of WO_3-ZrO_2 catalysts. To know the structure of zirconia, some structural models have been developed, which clarify that higher acidity is obtained when WO_x is supported on $ZrO_x(OH)_{4-2x}$ than when it is on the crystalline structure of ZrO_2. Then, tungsten oxides combined with Pt become active and selective catalysts for the hydroisomerization reaction of light n-alkanes.[54] However, the mechanism of the isomerization reaction with Pt/WO_x-ZrO_2 is still not elucidated:

Theory 1: Suggests that in the hydrogenation reaction, the formation of reactive intermediates (alkene) first occurs; this is followed by isomerization through a mononuclear reaction saturating the surface WO_x to reach surface saturation, while Pt with hydrogen flux promotes the reduction of surface WO_x and acts to remove coke.[55,56]

Theory 2. Suggests that in hydroisomerization with Pt/WO_x-ZrO_2, platinum sites are critical for the dissociation and storage of hydrogen required for the transfer and desorption of alkene ions.[50–56]

4.6   PROCESS VARIABLES

For the isomerization/hydroisomerization reaction to occur properly, there must be adequate control of conditions, that is before and during the isomerization, and variables such as:

Temperature: This is the most important of the operating variables. It depends on the chemical characteristics of the catalyst, since, as mentioned above, catalysts require a sufficient temperature to activate the catalytic material. Desulfurization reactions are favored at temperatures between 230°C and 340°C, such that nitrogen and oxygen compounds are lost and cracking reactions start at 350°C and above.

Pressure: This variable is also a function of the catalyst, as it must be sufficient to hold the molecules together. It has a direct effect on the removal of impurities; high pressure in the reactor allows the catalyst to last longer by avoiding the formation of carbon.

Catalyst: The chemical nature of the catalyst is particularly important, as it must have the ability to reduce the activation energy so that the molecules can reorganize themselves. The choice of the catalyst depends on the desired product and process conditions.

Hydrogen/hydrocarbon ratio: This is the cubic meters of hydrogen present in the system/cubic meters of hydrocarbon to be reacted. Hydrogen maintains physical contact between the hydrocarbon and the catalyst and ensures that chemical reactions occur at the active sites of the catalyst. Excess hydrogen prevents the formation of carbon on the catalyst. The ratio of hydrogen to hydrocarbon determines the partial pressure in the reactor.

4.7 PROCESS TECHNOLOGIES

The Axes isomerization process is a technology that includes the IPsorb and Hexorb processes, both of which are based on a chlorinated alumina catalyst.

The Hexorb process uses isohexane to desorb normal paraffin, while Ipsorb generates an isomerate with an octane number of 89.

The British Petroleum Company (BP) uses a Pt/alumina catalyst with carbon tetrachloride, with the advantage of single-stage operation.

The PENEX process developed by Universal Oil Products Co. (OUP) is based on the Friedel–Crafts reaction.

Innovations include zeolite-based catalysts, which are more resistant to deactivation but have low activity and selectivity.

The fifth generation consists of metal superalloys, such as specifically sulfonated zirconia and Pt-tungsten zirconia.

The catalyst based on Pt/SO_4^{-2}-ZrO_2 forms the basis of the UOP Par-Isom process (disadvantage, total degradation of the material upon loss of sulfur). They have shown high catalytic activity and water resistance.

The light naphtha isomerization process operated with a Pt/SO_4^{-2}-ZrO_2 catalyst developed by Cosmo Oil, Ltd, Mitsubishi Heavy Industries, Ltd, and UOP LLC occurred in 1996.

WO_x-ZrO_2 is the catalyst for Exxon's EMICT process. It has high conversion to isoparaffins, due to the low process temperature (thermodynamic) and minimizes decomposition losses. Both technologies use superacids operated at low temperatures, resulting in high selectivity to branched isomers.

In conclusion, the isomerization of light paraffin is important in the oil refining industry to produce clean and high-octane fuels. However, there are still challenges and limitations to the process, such as high temperatures and pressures; therefore, it has a high energy demand and is very costly. It should also be considered that the presence of impurities and contaminants in the reactor feed stream significantly affects the efficiency and selectivity, favoring undesirable side reactions. Despite these drawbacks, isomerization is essential for the oil industry, and with continuous advances, it is expected to reach more efficient and cost-effective technologies

The reaction with medium exothermicity takes place in acidic conditions and, from a thermodynamic point of view, it is favored at low temperatures because this is when the branching of the structural chain is reached. Although it must be remembered that this process is limited to n-alkanes and iso-alkanes of low molecular weight, the process with solid acidic catalysts implies high temperatures and pressures, high economic costs for the production of the catalyst, and high replacement costs in the short term, operating costs, etc.

4.8 CONCLUSIONS AND OUTLOOK

As described above, isomerization is a core process in hydrocarbon refining, and it indicates that this trend will continue for several decades as it represents a solution to the increasing demand for high-octane gasoline. The growing demand for clean fuels stimulates the development of more efficient and sustainable methods. Although great progress has been made with the implementation of new technologies, the isomerization process should continue to improve and make the production of isomers more efficient.

Nanoscale catalysts have shown good performance and represent a promising technology with significant improvements in selectivity.

The desirability of venturing into the isomerization of biofuels is attractive to improve their properties and achieve more sustainable and efficient products.

REFERENCES

(1) Gary, J. H.; Handwerk, G. E.; Kaiser, M. J. *Petroleum Refining: Technology and Economics*, 5th Edition; CRC Press, 2007.

(2) Biermann, U.; Metzger, J. O. Synthesis of Alkyl-branched Fatty Acids. *Eur. J. Lipid Sci. Technol.* 2008, *110* (9), 805–811. https://doi.org/10.1002/ejlt.200800033

(3) Ono, Y. A Survey of the Mechanism in Catalytic Isomerization of Alkanes. *Catal. Today* 2003, *81* (1), 3–16. https://doi.org/10.1016/S0920-5861(03)00097-X

(4) Kaunisto, K.; Lagerbom, J.; Honkanen, M.; Varis, T.; Lambai, A.; Mohanty, G.; Levänen, E.; Kivikytö-Reponen, P.; Frankberg, E. Evolution of Alumina Phase Structure in Thermal Plasma Processing. *Ceram. Int.* 2023, *49* (13), 21346–21354. https://doi.org/10.1016/j.ceramint.2023.03.263

(5) Oudejans, J. C. Zeolite Catalysts in Some Organic Reactions, 1984. https://repository.tudelft.nl/islandora/object/uuid%3A8d81fbee-fc54-4698-9587-e12c7fe51de1

(6) Chen, N. Y.; Garwood, W. E. Industrial Application of Shape-Selective Catalysis. *Catal. Rev.* 1986, *28* (2–3), 185–264. https://doi.org/10.1080/01614948608082251

(7) Loeffler, E.; Lohse, U.; Peuker, Ch.; Oehlmann, G.; Kustov, L. M.; Zholobenko, V. L.; Kazansky, V. B. Study of Different States of Nonframework Aluminum in Hydrothermally Dealuminated HZSM-5 Zeolites Using Diffuse Reflectance i.r Spectroscopy. *Zeolites* 1990, *10* (4), 266–271.

(8) Dejaifve, P.; Auroux, A.; Gravelle, P. C.; Védrine, J. C.; Gabelica, Z.; Derouane, E. G. Methanol Conversion on Acidic ZSM-5, Offretite, and Mordenite Zeolites: A Comparative Study of the Formation and Stability of Coke Deposits. *J. Catal.* 1981, *70* (1), 123–136. https://doi.org/10.1016/0021-9517(81)90322-5

(9) Leyva, C.; Rana, M. S.; Trejo, F.; Ancheyta, J. On the Use of Acid-Base-Supported Catalysts for Hydroprocessing of Heavy Petroleum. *Ind. Eng. Chem. Res.* 2007, *46* (23), 7448–7466. https://doi.org/10.1021/ie070128q

(10) Dorado, F.; Romero, R.; Cañizares, P. Influence of Clay Binders on the Performance of Pd/HZSM-5 Catalysts for the Hydroisomerization of n-Butane. *Ind. Eng. Chem. Res.* 2001, *40* (16), 3428–3434. https://doi.org/10.1021/ie001133w

(11) Cañizares, P.; Dorado, F.; Sánchez, P.; Romero, R. Hydroisomerization of N-Butane over Pd/HZSM-5 and Pd/Hmordenite with and without Binder. *Stud. Surf. Sci. Catal.* 2002, *142*, 707–714. https://doi.org/10.1016/S0167-2991(02)80092-1.

(12) Tamizhdurai, P.; Ramesh, A.; Krishnan, P. S.; Narayanan, S.; Shanthi, K.; Sivasanker, S. Effect of Acidity and Porosity Changes of Dealuminated Mordenite on N-Pentane, n-Hexane and Light Naphtha Isomerization. *Micropor. Mesopor. Mat.* 2019, *287*, 192–202. https://doi.org/10.1016/j.micromeso.2019.06.012

(13) Al-Rawi, U. A.; Sher, F.; Hazafa, A.; Bilal, M.; Lima, E. C.; Al-Shara, N. K.; Jubeen, F.; Shanshool, J. Synthesis of Zeolite Supported Bimetallic Catalyst and Application in N-Hexane Hydro-Isomerization Using Supercritical CO_2. *J. Environ. Chem. Eng.* 2021, *9* (4), 105206. https://doi.org/10.1016/j.jece.2021.105206

(14) V.P. Kukhar Institute of Bioorganic Chemistry and Petrochemistry of National Academy of Sciences of Ukraine 1, Murmans'ka St., 02660 Kyiv, Ukraine; Patrylak, L.; Krylova, M.; N-Hexane Isomerization Over Nickel-Containing Mordenite Zeolite. *Chem. Chem. Technol.* 2020, *14* (2), 234–238. https://doi.org/10.23939/chcht14.02.234

(15) Li, T.; Wang, W.; Feng, Z.; Bai, X.; Su, X.; Yang, L.; Jia, G.; Guo, C.; Wu, W. The Hydroisomerization of *n*-Hexane over Highly Selective Pd/ZSM-22 Bifunctional Catalysts: The Improvements of Metal-Acid Balance by Room Temperature Electron Reduction Method. *Fuel* 2020, *272*, 117717. https://doi.org/10.1016/j.fuel.2020.117717

(16) Vajglová, Z.; Kumar, N.; Peurla, M.; Hupa, L.; Semikin, K.; Sladkovskiy, D. A.; Murzin, D. Yu. Effect of the Preparation of Pt-Modified Zeolite Beta-Bentonite Extrudates on Their Catalytic Behavior in *n* -Hexane Hydroisomerization. *Ind. Eng. Chem. Res.* 2019, *58* (25), 10875–10885. https://doi.org/10.1021/acs.iecr.9b01931

(17) Van Der Wal, L. I.; Oenema, J.; Smulders, L. C. J.; Samplonius, N. J.; Nandpersad, K. R.; Zečević, J.; De Jong, K. P. Control and Impact of Metal Loading Heterogeneities at the Nanoscale on the Performance of Pt/Zeolite Y Catalysts for Alkane Hydroconversion. *ACS Catal.* 2021, *11* (7), 3842–3855. https://doi.org/10.1021/acscatal.1c00211

(18) Eswaramoorthi, I.; Lingappan, N. Hydroisomerisation of N-Hexane over Bimetallic Bifunctional Silicoaluminophosphate Based Molecular Sieves. *Appl. Catal. Gen.* 2003, *245*, 119–135. https://doi.org/10.1016/S0926-860X(02)00637-3

(19) Kang, Y.; Rao, X.; Yuan, P.; Wang, C.; Wang, T.; Yue, Y. Al-Functionalized Mesoporous SBA-15 with Enhanced Acidity for Hydroisomerization of n-Octane. *Fuel Process. Technol.* 2021, *215*, 106765. https://doi.org/10.1016/j.fuproc.2021.106765

(20) Vermeiren, W.; Gilson, J.-P. Impact of Zeolites on the Petroleum and Petrochemical Industry. *Top. Catal.* 2009, *52* (9), 1131–1161. https://doi.org/10.1007/s11244-009-9271-8

(21) Song, X.; Sayari, A. Sulfated Zirconia-Based Strong Solid-Acid Catalysts: Recent Progress. *Catal. Rev.* 1996, *38* (3), 329–412. https://doi.org/10.1080/01614949608006462

(22) Kanougi, T.; Atoguchi, T.; Yao, S. Periodic Density Functional Study of Superacidity of Sulfated Zirconia. *J. Mol. Catal. Chem.* 2002, *177* (2), 289–298. https://doi.org/10.1016/S1381-1169(01)00276-X

(23) Vernon, C. F. and Grant, C. B., Bartlesville, O. *Sulfate-treated zirconia-gel catalyst.* Phillips Petroleum Company. 1962. US3032599.

(24) Adeeva, V.; Liu, H.-Y.; Xu, B.-Q.; Sachtler, W. M. H. Alkane Isomerization over Sulfated Zirconia and Other Solid Acids. *Top. Catal.* 1998, *6* (1), 61–76. https://doi.org/10.1023/A:1019114406219

(25) Zalewski, D. J.; Alerasool, S.; Doolin, P. K. Characterization of Catalytically Active Sulfated Zirconia. *Catal. Today* 1999, *53* (3), 419–432. https://doi.org/10.1016/S0920-5861(99)00137-6

(26) Reddy, B. M.; Patil, M. K. Organic Syntheses and Transformations Catalyzed by Sulfated Zirconia. *Chem. Rev.* 2009, *109* (6), 2185–2208. https://doi.org/10.1021/cr900008m

(27) Jentoft, F. C.; Hahn, A.; Kröhnert, J.; Lorenz, G.; Jentoft, R. E.; Ressler, T.; Wild, U.; Schlögl, R.; Häßner, C.; Köhler, K. Incorporation of Manganese and Iron into the Zirconia Lattice in Promoted Sulfated Zirconia Catalysts. *J. Catal.* 2004, *224* (1), 124–137. https://doi.org/10.1016/j.jcat.2004.02.012

(28) Paál, Z.; Wild, U.; Muhler, M.; Manoli, J.-M.; Potvin, C.; Buchholz, T.; Sprenger, S.; Resofszki, G. The Possible Reasons of Irreversible Deactivation of Pt/Sulfated Zirconia Catalysts: Structural and Surface Analysis. *Appl. Catal. Gen.* 1999, *188* (1), 257–266. https://doi.org/10.1016/S0926-860X(99)00211-2

(29) Morterra, C.; Cerrato, G.; Bolis, V. Lewis and Brønsted Acidity at the Surface of Sulfate-Doped ZrO_2 Catalysts. *Catal. Today* 1993, *17* (3), 505–515. https://doi.org/10.1016/0920-5861(93)80053-4

(30) Brown, A. S. C.; Hargreaves, J. S. J. Sulfated Metal Oxide Catalysts. *Green Chem.* 1999, *1* (1), 17–20. https://doi.org/10.1039/a807963c

(31) Kustov L.M., Kazansky V.B., Figueras F., Tichit D. Investigation of the Acidic Properties of ZrO2 Modified by SO_2–4 Anions. *J. Catal.* 1994, 143–149. https://doi.org/10.1006/jcat.1994.1330

(32) Song, X.; Reddy, K. R.; Sayari, A. Effect of Pt and H_2O on n-Butane Isomerization over Fe and Mn Promoted Sulfated Zirconia. *J. Catal.* 1996, *161* (1), 206–210. https://doi.org/10.1006/jcat.1996.0178

(33) Fals, J.; García, J. R.; Falco, M.; Sedran, U. Coke from SARA Fractions in VGO. Impact on Y Zeolite Acidity and Physical Properties. *Fuel* 2018, *225*, 26–34. https://doi.org/10.1016/j.fuel.2018.02.180

(34) Sadrameli, S. M. Thermal/Catalytic Cracking of Liquid Hydrocarbons for the Production of Olefins: A State-of-the-Art Review II: Catalytic Cracking Review. *Fuel* 2016, *173*, 285–297. https://doi.org/10.1016/j.fuel.2016.01.047

(35) Li, X.; Nagaoka, K.; Simon, L. J.; Olindo, R.; Lercher, J. A. Mechanism of Butane Skeletal Isomerization on Sulfated Zirconia. *J. Catal.* 2005, *232* (2), 456–466. https://doi.org/10.1016/j.jcat.2005.03.025

(36) Hino, M.; Arata, K. Synthesis of Solid Superacid of Tungsten Oxide Supported on Zirconia and Its Catalytic Action for Reactions of Butane and Pentane. *J. Chem. Soc. Chem. Commun.* 1988, No. 18, 1259–1260. https://doi.org/10.1039/C39880001259

(37) Hidalgo, J. M.; Kaucký, D.; Bortnovsky, O.; Černý, R.; Sobalík, Z. Isomerization of C5–C7 Paraffins over a Pt/WO_3–ZrO_2 Catalyst Using Industrial Feedstock. *Monatshefte Für Chem. - Chem. Mon.* 2014, *145* (9), 1407–1416. https://doi.org/10.1007/s00706-014-1231-8

(38) Smolikov, M. D.; Shkurenok, V. A.; Dzhikiya, O. V.; Kir'yanov, D. I.; Zatolokina, E. V.; Belyi, A.S. Features of the Phase Composition, Acidity, and Isomerization Activity of $Pt/WO_3/ZrO_2$ Catalysts. *Russ. Chem. Bull.* 2020, *69* (9), 1714–1718. https://doi.org/10.1007/s11172-020-2953-x

(39) Ciptonugroho, W.; Al-Shaal, M. G.; Mensah, J. B.; Palkovits, R. One Pot Synthesis of WO*x*/Mesoporous-ZrO_2 Catalysts for the Production of Levulinic-Acid Esters. *J. Catal.* 2016, *340*, 17–29. https://doi.org/10.1016/j.jcat.2016.05.001

(40) Barrera, A.; Montoya, J. A.; Viniegra, M.; Navarrete, J.; Espinosa, G.; Vargas, A.; del Angel, P.; Pérez, G. Isomerization of *n*-Hexane over Mono- and Bimetallic Pd–Pt Catalysts Supported on ZrO_2–Al_2O_3–WOx Prepared by Sol–Gel. *Appl. Catal. Gen.* 2005, *290* (1), 97–109. https://doi.org/10.1016/j.apcata.2005.05.011

(41) Cortés-Jácome, M. A.; Angeles-Chavez, C.; López-Salinas, E.; Navarrete, J.; Toribio, P.; Toledo, J. A. Migration and Oxidation of Tungsten Species at the Origin of Acidity and Catalytic Activity on WO_3-ZrO_2 Catalysts. *Appl. Catal. Gen.* 2007, *318*, 178–189. https://doi.org/10.1016/j.apcata.2006.11.019

(42) Hernández, M. L.; Montoya, J. A.; Del Angel, P.; Hernández, I.; Espinosa, G.; Llanos, M. E. Influence of the Synthesis Method on the Nanostructure and Reactivity of Mesoporous Pt/Mn-WOx-ZrO_2 Catalysts. *Catal. Today* 2006, *116* (2), 169–178. https://doi.org/10.1016/j.cattod.2006.02.084

(43) He, G.; Zhang, R.; Zhao, Q.; Yang, S.; Jin, H.; Guo, X. Effect of the Cr_2O_3 Promoter on Pt/WO_3-ZrO_2 Catalysts for n-Heptane Isomerization. *Catalysts* 2018, *8* (11), 522. https://doi.org/10.3390/catal8110522

(44) Smolikov, M. D.; Shkurenok, V. A.; Kir'yanov, D. I.; Belyi, A. S. Active Surface Formation of Tungstated Zirconia Catalysts for N-Heptane Isomerization. *Catal. Today* 2019, *329*, 63–70. https://doi.org/10.1016/j.cattod.2019.01.036

(45) Piva, D. H.; Piva, R. H.; Pereira, C. A.; Silva, D. S. A.; Montedo, O. R. K.; Morelli, M. R.; Urquieta-González, E. A. Facile Synthesis of WOx/ZrO_2 Catalysts Using WO_3·H_2O Precipitate as Synthetic Precursor of Active Tungsten Species. *Mater. Today Chem.* 2020, *18*, 100367. https://doi.org/10.1016/j.mtchem.2020.100367

(46) Martínez, A.; Prieto, G.; Arribas, M. A.; Concepción, P.; Sánchez-Royo, J. F. Influence of the Preparative Route on the Properties of WOx–ZrO_2 Catalysts: A Detailed Structural, Spectroscopic, and Catalytic Study. *J. Catal.* 2007, *248* (2), 288–302. https://doi.org/10.1016/j.jcat.2007.03.022

(47) Scheithauer, M.; Grasselli, R. K.; Knözinger, H. Genesis and Structure of WOx/ZrO_2 Solid Acid Catalysts. *Langmuir* 1998, *14* (11), 3019–3029. https://doi.org/10.1021/la971399g

(48) Baertsch, C. D.; Komala, K. T.; Chua, Y.-H.; Iglesia, E. Genesis of Brønsted Acid Sites during Dehydration of 2-Butanol on Tungsten Oxide Catalysts. *J. Catal.* 2002, *205* (1), 44–57. https://doi.org/10.1006/jcat.2001.3426

(49) *Nature of Catalytically Active Sites in the Supported WO_3/ZrO_2 Solid Acid System: A Current Perspective | ACS Catalysis.* https://pubs.acs.org/doi/abs/10.1021/acscatal.6b03697 (accessed August 05, 2024.

(50) Barton, D. G.; Soled, S. L.; Meitzner, G. D.; Fuentes, G. A.; Iglesia, E. Structural and Catalytic Characterization of Solid Acids Based on Zirconia Modified by Tungsten Oxide. *J. Catal.* 1999, *181* (1), 57–72. https://doi.org/10.1006/jcat.1998.2269

(51) Yue, C.; Zhu, X.; Rigutto, M.; Hensen, E. Acid Catalytic Properties of Reduced Tungsten and Niobium-Tungsten Oxides. *Appl. Catal. B Environ.* 2015, *163*, 370–381. https://doi.org/10.1016/j.apcatb.2014.08.008

(52) Wong, S.-T.; Li, T.; Cheng, S.; Lee, J.-F.; Mou, C.-Y. Aluminum-Promoted Tungstated Zirconia Catalyst in *n*-Butane Isomerization Reaction. *J. Catal.* 2003, *215* (1), 45–56. https://doi.org/10.1016/S0021-9517(02)00176-8

(53) Bigey, C.; Hilaire, L.; Maire, G. WO_3–CeO_2 and Pd/WO_3–CeO_2 as Potential Catalysts for Reforming Applications. *J. Catal.* 2001, *198* (2), 208–222. https://doi.org/10.1006/jcat.2000.3111

(54) Filimonova, S. V.; Nosov, A. V.; Scheithauer, M.; Knözinger, H. *N*-Pentane Isomerization over Pt/WOx/ZrO_2 Catalysts: A 1H and 13C NMR Study. *J. Catal.* 2001, *198* (1), 89–96. https://doi.org/10.1006/jcat.2000.3112

(55) Kuba, S.; Lukinskas, P.; Ahmad, R.; Jentoft, F. C.; Grasselli, R. K.; Gates, B. C.; Knözinger, H. Reaction Pathways in *n*-Pentane Conversion Catalyzed by Tungstated Zirconia: Effects of Platinum in the Catalyst and Hydrogen in the Feed. *J. Catal.* 2003, *219* (2), 376–388. https://doi.org/10.1016/S0021-9517(03)00233-1

(56) Iglesia, E.; Barton, D. G.; Soled, S. L.; Miseo, S.; Baumgartner, J. E.; Gates, W. E.; Fuentes, G. A.; Meitzner, G. D. Selective Isomerization of Alkanes on Supported Tungsten Oxide Acids. In *Studies in Surface Science and Catalysis*; Elsevier, 1996; Vol. 101, pp 533–542. https://doi.org/10.1016/S0167-2991(96)80264-3

5 Alkylate Gasoline

Ricardo Cerón-Camacho and
Diego Guzmán-Lucero
Mexican Petroleum Institute, Mexico City, México

5.1 INTRODUCTION

The transportation of goods and people worldwide is an energy-intensive activity that mainly involves light and freight vehicles[1] and is correlated with population growth and the access of societies to more consumer goods. Its increase is expected to be more pronounced in non-OECD (Organization for Economic Co-operation and Development) countries, mainly India and China, reaching 2 billion light-duty cars and almost 800 million freight vehicles globally by 2040. From 2025 to 2040, air travel is projected to be the fastest-growing means of transportation, which is highly dependent on gross domestic product (GDP) growth.[2] It is estimated that 20% of the consumed energy corresponds to transportation and contributes to 14% of greenhouse gases (GHGs).[3] On the other hand, approximately 80% of cars use spark ignition engines. In contrast, diesel engines predominate in the commercial sector,[4] and so about 60% of crude oil is used to make transport fuels, of which almost 5 billion liters per day correspond to gasoline.[5,6] In this context, in many countries, stricter legislation, including fuel taxes, carbon emission taxes, and subsidies for renewable energies, has been implemented to reduce pollutant emissions.[7,8]

TABLE 5.1
Properties of Industrial Liquid Acid Catalysts

Properties	HF	H_2SO_4
Molecular weight, g/gmol	20.0	98.08
Density, g/mL	1.002 (4 °C)	1.8 (20 °C)
Boiling point, °C (760 mm Hg)	19.5	290
Freezing point, °C	−83	10.31
Vapor pressure, mm Hg	917 (25 °C)	1 (146 °C)
Viscosity, cP	0.256 (0 °C)	33 (15 °C) (21 mPa.s (25 °C))
Surface tension, mN/m	10.2 (0 °C)	51.7 (50 °C)
Specific heat, Btu/lb/°F	0.83 (−1 °C)	0.33 (20 °C)
Hammet acidity (-H_0), 25 °C	10.0	11.1
Dielectric constant	84 (0 °C)	114 (20 °C)

DOI: 10.1201/9781003517283-5

This chapter has been made available under a CC-BY-NC-ND 4.0 license.

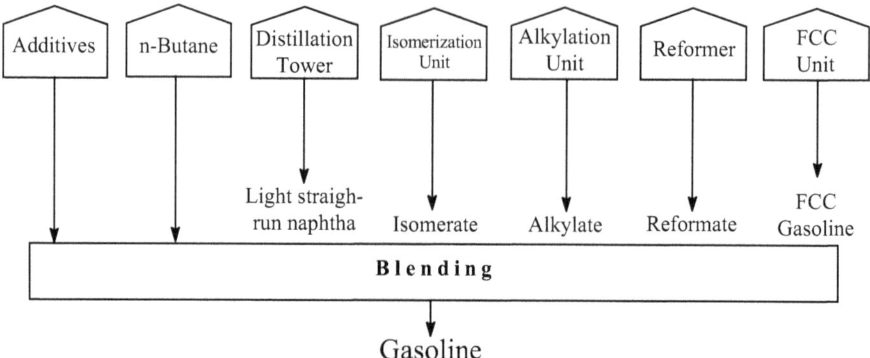

FIGURE 5.1 General outline of gasoline production.

Gasoline is a complex mixture of hydrocarbons that also depends on the time and place where it is produced[8] and involves the use of different and complex technologies to produce it as shown in Figure 5.1.[9] It is a volatile and flammable mixture of hydrocarbons obtained mainly from petroleum refining.[10] The gasoline pool is composed of streams coming from different units, including straight-run naphtha, isomerate, reformate, alkylate, fluidized catalytic cracking (FCC), hydrocracked naphtha, and polymer gasoline,[11,12] and contains compounds such as alkanes (paraffins), isoalkanes (isoparaffins), olefins, naphthenes, and aromatics, known as PIONA.[13] Gasoline consists of different proportions of these streams, and other special additives must meet ecological and quality requirements to function appropriately in engines and, consequently, generate low exhaust emissions. Therefore, the composition and properties of fuels are guided by legal provisions.[14] The streams that makeup gasoline are at boiling points ranging from −20 to 230 °C, and its components have a carbon number distribution between C4 and C12. An accelerated decrease in gasoline demand is expected, mainly in OECD countries due to the implementation of decarbonization and technologies with lower CO_2 emissions; in contrast, developing countries still show a tendency to increase consumption until 2050 because they are societies that do not have a saturated automotive market. However, in developed countries, the number of cars per 1000 inhabitants has displayed a constant trend (e.g., in the USA, 838 and 836 in 2008 and 2018, respectively), but in developing countries, the number of cars per 1000 inhabitants has a wide margin to grow (e.g., India 13.2 and 43.6, China 35.7 and 167, or Mexico 230.2 and 343, for the years 2008 and 2018, respectively).[8,11,15]

As for cars, knocking in gasoline engines is caused by the rapid combustion of an air–fuel mixture that does not burn in the regular flame front. When air is compressed with gasoline, the latter has to self-ignite before the ignition spark, thus creating resistance to the engine compression stroke, which is known as knocking. Moreover, adequate engine fuel efficiency is an essential characteristic. In this context, the anti-knock performance is measured from the octane number (ON), which is one of the essential properties of gasoline and probably the most recognized measurement of the quality of gasoline and is defined as the volume percentage of *i*-octane in a mixture of n-heptane and i-octane.[16] There are two ASTM methods for measuring this index: the Research Octane Number (RON, ASTM D2699) and the Motor Octane Number (MON, ASTM D2700). Always RON > MON, and the difference between them is known as gasoline sensitivity.[17] Historically, the ON has been increased to decrease the antiknock performance. Sulfur and nitrogen, among others, have been decreased to reduce the environmental impact of the output emissions, which considerably decreased the contents of hydrocarbons, CO, NO_x, SO_x, and particulate materials, whereas the presence of oxygenated compounds has been increased.[12,18] From the point of view of using fuels from nonrenewable sources, not only the gasoline pool has been improved but also the efficiency of the engines has decreased due to the impact of emissions and increasing digitalization.[19] Car manufacturing companies have developed and employed technologies aimed at reducing specific fuel consumption in engines and achieving greater efficiency[6,20] that include continuously variable transmission, cylinder deactivation, start-stop systems, variable valve timing, direct fuel injection, and turbocharging technologies.[8] The use of electronic control generated the possibility of increasing the compression ratio of engines (from 8 in 1970 to 9.5 in 2005), which resulted in less-polluting exhaust gases that allowed the implementation of Euro and Tier standards.[8]

Alkylation is an essential refining process for producing higher-quality gasoline and meeting increasingly stringent standards for air quality, carbon emissions, and automotive efficiency. Due to alkylation, carbon emissions have been decreased by half, and emissions of air pollutants have fallen by 80% since the 1970s.[21] The combination of the different streams that make up the gasoline pool (Figure 5.1) and other components, such as ethanol, made it possible to increase the octane rating, which has helped the engine performance achieve lower GHG emissions. Alkylate provides other attributes that reduce environmental pollution such as low volatility and low evaporative emissions, resulting in low Reid vapor pressure (RVP), high RON and MON values, and no aromatic or sulfur compounds.[10] Because of the above characteristics, the demand for the contribution of the other streams to the gasoline pool was reduced, except for the alkylate.[22]

The development of the alkylation reaction dates back to the 1930s when companies such as UOP, Shell, Anglo Iranian Oil Company, and Texaco made public alkylate obtainments. The employed catalysts were combinations of strong Lewis acids ($AlCl_3$ or BF_3) with acids such as HCl or hydrofluoric acid (HF).[23,24] In 1938, the first commercial alkylation plant, which used sulfuric acid as a catalyst, was built in Texas by the Humble Oil Company. The technology featuring sulfuric acid was dominant until the mid-1970s (75%) when HF catalysis became more important until an alkylate yield similar to that was obtained with H_2SO_4.[25] Initially, alkylate production was focused on meeting the demand for high-octane fuels for allied aviation during World War II; the catalysts in the alkylation plants were H_2SO_4 and HF.[26,27] Starting in 1950, the production of alkylate gasoline for automobile engines began; however, its use still needed to be improved because this component of the gasoline pool was expensive.[28] With the continuous enactment of stricter legislation for environmental care, the alkylate component of the gasoline pool has become highly relevant, and currently, premium gasoline can reach up to 60% of the content.[27] Nowadays, most industrial processes use sulfuric acid as a catalyst because it is more economical, safer to handle, and more environmentally friendly than HF.[29]

Almost all alkylates are obtained by using sulfuric acid or HF. The license holders that are used to produce alkylates employing H_2SO_4 are ExxonMobil (auto cooling process) and DuPont STRATCO (Stratford Engineering Corporation) (flow cooling process). Currently, the only one with the license to obtain alkylate by means of HF is UOP LLC.[27] There are other technologies with which alkylate is produced, such as Honeywell UOP, which commercialized the ISOALKY process in 2016, an alkylation technology developed by Chevron USA, and the Ionikylation technology developed by PetroChina, which installed its first commercial unit in 2013. Both technologies use ionic liquids (ILs) in the production of alkylates. In 2015, the first unit using a solid acid catalyst was installed with Albemarle AlkyStar technology, and the catalyst was based on zeolites.

Therefore, this chapter will review the commercial alkylation processes using H_2SO_4 and HF acidic liquid catalysts and the current developments using IL catalysts and solid catalysts such as zeolites.

5.2 ALKYLATION USING STRONG ACIDS

5.2.1 SULFURIC ACID

Concentrated sulfuric acid is used as a liquid catalyst to produce alkylate gasoline by reacting isobutane with olefins such as propylene, butylenes, or amylenes (C3–C5 olefins). Butylenes are the primary raw materials for obtaining the highest quality alkylates. The result is highly branched (C7–C9) hydrocarbons, mainly isooctanes, which have the highest RON (90–98) and MON (88–95) number values. In practice, alkylate is a complex mixture of isoparaffins and other hydrocarbons.[30] Until 1996, the most widely used acid catalyst was HF. Due to its high toxicity (when leaked, it can cause dangerous and stable aerosols[31,32]), more emphasis has been placed on the development of sulfuric acid-based technology and, in many industrialized countries, licenses for the construction of new HF-based alkylation plants are no longer available.[33] Alkylate is a premium component of the gasoline pool because it has outstanding antiknock properties and can combust cleanly.[34] As expected, using a catalyst in the olefin reactions (C3–C5) and isobutane allowed milder reaction conditions, temperatures below 50°C and pressures below 30 bars. The alkylate quality depends on factors such as the initial olefin mixture, isobutane-to-olefin ratio, H_2SO_4-to-reactants ratio, and stirring efficiency.[35,36]

The olefins fed to the alkylation unit generally come from an FCC unit and the main components are butene and isobutene and minor amounts of propane and amylenes. Water is removed from this stream. Isobutane comes from the catalytic reforming unit stream (70% vol.) and the high purity (over 95% vol.) comes from the alkylation unit of the deisobutanizer tower.

In the STRATCO effluent-cooled process (Figure 5.2), olefins, isobutane, and recycled acid are introduced into the reactor (a continuous flow stirred tank) and emulsified using a high-power mixer located at one end of the reactor for 20–35 min. The hydrocarbon mixture with the acid is circulated through the heat transfer tubes

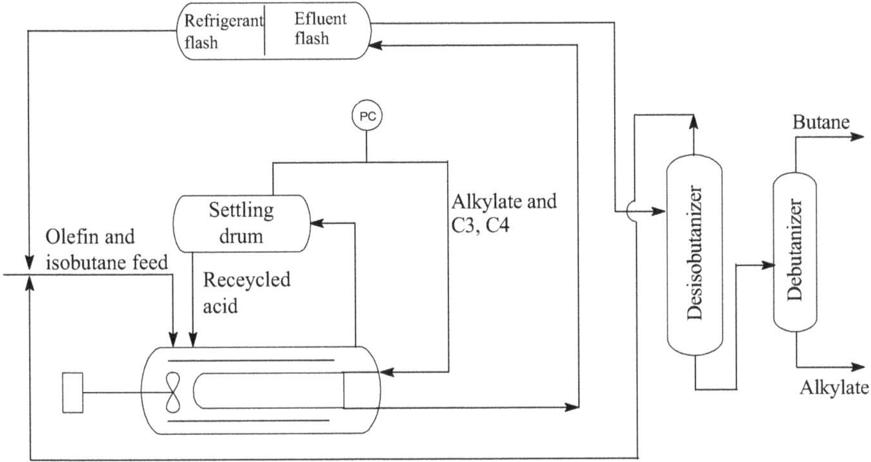

FIGURE 5.2 Alkylation unit of the refrigerated STRATCO effluent process.

repeatedly. Once the reaction is complete, the emulsion passes into a decanter, where the acid is separated by gravity from the hydrocarbons for 30–60 min. The separated hydrocarbon stream may contain approximately 80 wt. % of isobutane. This stream is reduced in pressure so that the light hydrocarbons evaporate (flash) and the temperature of the liquid hydrocarbons decreases, which is used as the cooling stream in the reactor. Most of the acid is recycled to the reactor and the portion of acid that is not recycled is replaced by fresh acid (98%). The acid concentration was kept above 90% for optimum selectivity and activity. The temperature in the reactor depends on the composition of the olefin feed; when the feed stream has butenes and pentenes, it is between 1 and 5°C and when it is a propylene-rich stream, it is 10°C.

In ExxonMobil's self-cooling cascade alkylation process (Figure 5.3), olefins, acid, and isobutane, which also act as a coolant, are introduced into the first zone of the cascade reactor. In the next reactor zones, isobutane/olefin mixtures are injected

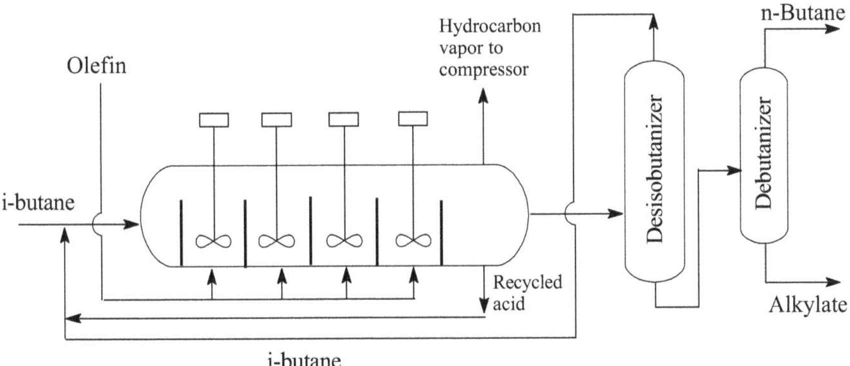

FIGURE 5.3 Alkylation unit of the autorefrigeration ExxonMobil process.

at ratios ranging from 8:1 to 12:1. Each zone of the cascade reaction enters an emulsion of acid and hydrocarbons from the previous zone. Starting from the second reaction section, the hydrocarbon in the emulsion is mainly composed of isobutane and alkylate. In the first reaction zone, the lowest temperature (1.2°C) and higher concentration of isobutane are recorded and the best alkylate is obtained. As one moves to the next reaction zone, the temperature increases (up to 7°C), and because the concentration of olefins introduced into each reaction zone increases, the quality of the alkylate decreases.[37] The evaporating gases are compressed and sent to the feed, which allows the cooling of the feed stream. After the last reaction zone is allowed to settle, the acid phase is pumped to the feed and the hydrocarbon phase in the treatment zone of these effluents (mainly isobutane) is recovered in the deisobutanizer to recycle it again.

In hydrocarbon streams, sulfate esters are obtained, which are removed by caustic washing. An alternative is acid washing, which unlike caustic washing, can produce more alkylate, regenerate the acid, and consume less acid.[38]

STRATCO reactors are designed to produce around 2,500 barrels of alkylate per day compared to ExxonMobil reactors, which are designed to produce between 3,000 and 10,000 barrels per day. The former allows greater operating flexibility. The significant cost associated with the alkylation process is due to the high sulfuric acid consumption (70–100 kg of acid per metric ton of alkylate) and the high energy consumption required to recover the deactivated acid (2–3 times more expensive than fresh acid).[39,40]

In industrial alkylation reactions, a large number of isoparaffins can be detected, between 130–150 (C5–C16);[41] many of the reaction products are isomers, so there is a complex distribution, and it is far beyond obtaining trimethylpentane (TMP) isomers (RON 100–109.6), which are the most desired products.[42] Other products such as C5–C7 (light ends) (RON 24.8–93), dimethylhexanes (DMH) (RON 55.5–73.6), methylheptanes (RON 21.7–26.8), and C9+ (heavy ends) (RON 70–91) are also obtained.[43]

The carbanion chain reaction mechanism mainly drives the reaction between *i*-butane and light olefins, which is catalyzed by concentrated sulfuric acid.[43,44] Olefins are converted to butyl carbocations by protonation of the double bonds to yield *sec*-butyl or *tert*-butyl cations[45,46] that can be dimerized or oligomerized with olefins to generate larger C8 to C12 carbocations. These larger carbocations can undergo cracking, resulting in light carbocations (C5–C7).[47,48] To complete the chain reactions, the hydride transferred between the TMP carbocation and isobutane forms TMP and an *i*-butyl cation (Figure 5.4).[49]

5.2.2 HYDROFLUORIC ACID

As well as sulfuric acid, the alkylation reactions using HF are important in organic chemistry for introducing an alkyl group to a substrate. Also, HF is another strongest Lewis acid that can activate electrophiles and nucleophiles, making it a versatile reagent for alkylation reactions. Table 5.1 describes the properties of HF, where the Hammett acidity value is lower than that of sulfuric acid.[26,50] However, it is important to note that HF is a very toxic and corrosive acid that can attack and corrode glass surfaces.

The alkylation reaction process for obtaining gasoline in petroleum refining is similar to that of using sulfuric acid[50]. Additionally, it produces high-octane gasoline

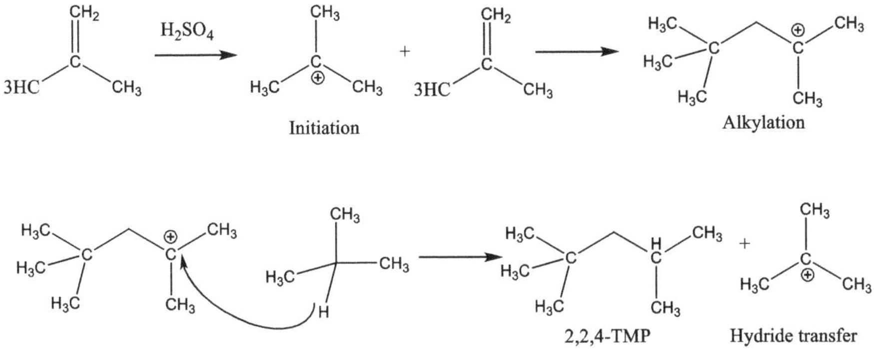

FIGURE 5.4 Simplified general scheme of the reaction mechanism between isobutylene and isobutane.

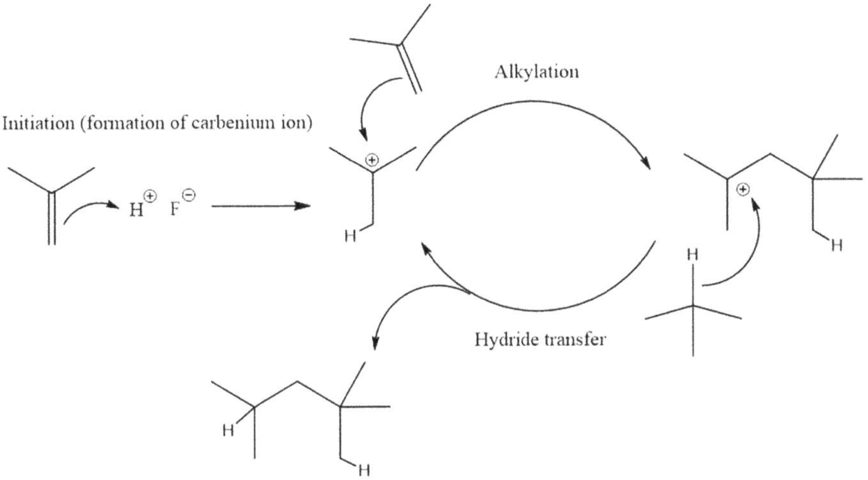

FIGURE 5.5 Alkylation mechanism with HF as catalyst.

components. It combines light and gaseous hydrocarbons (C3 to C5) to form heavier (C6 to C12)-branched chain hydrocarbons with excellent antiknock properties.

The mechanism of the alkylation reaction in HF is very well known and consists of the conversion of butenes and *i*-butenes to produce alkylate derivatives in three steps: the first one is the protonation of alkene derivatives to form carbenium ions, the second one is the alkylation reaction, and finally, hydride migration occurs to obtain the alkylate product, and the new carbenium species continue the catalytic cycle (Figure 5.5).[50]

As discussed below, strong acids are the best catalysts for alkylation reactions. However, the use of HF (although it is a better catalyst and industrially used) is associated with problems and serious challenges in terms of environmental aspects, corrosion problems in plants, and intrinsic handling danger.

In the refinery, when HF is used as a catalyst, the unit is called a HF alkylation unit (HFAU) and requires expensive corrosion-resistant materials for building the equipment. In general, the description of the HFAU is as follows: the first step of the reactor

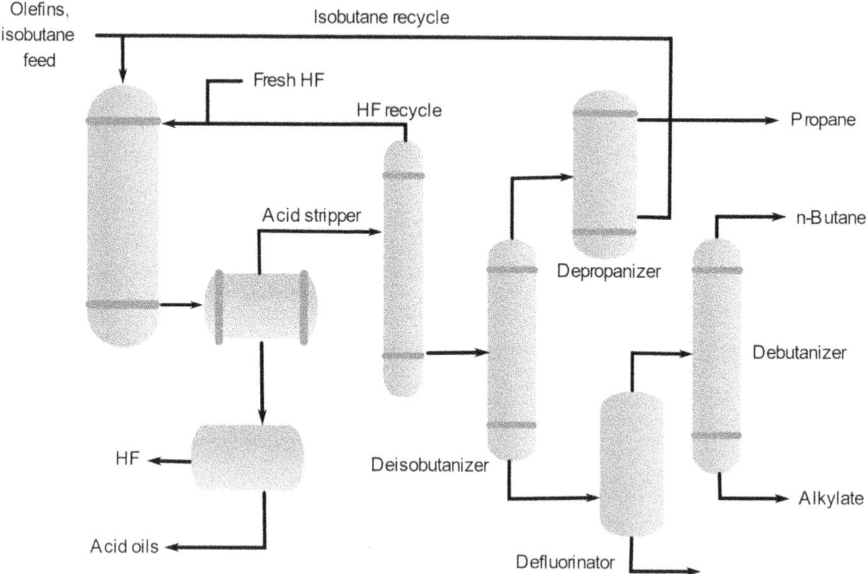

FIGURE 5.6 Schematization of the HFAU to obtain alkylate.

consists of olefin and isobutane feed treatment for coalescence and removal of water, sulfur, and other contaminants; later, the mixture of olefins and isobutene is passed to the reactor (with reaction chambers filled with HF); after the alkylation reaction occurs, the reactor effluent goes to a settler to separate the C7–C8 naphtha range product from the acid (this acid is separated and recirculated to the reactor); the alkylate fraction is washed in a caustic scrubber and later sent to a fractionation unit; in this last step, the fractionation consists of the deisobutanizer and depropanizer processes for obtaining the alkylate that is sent to gasoline pool blending. Figure 5.6 schematizes the whole described process. The operation conditions in the presence of sulfuric acid are discussed below and are similar to those in the presence of HF.[51,52]

It is important to highlight that the HFAU is provided mainly by UOP Honeywell at the industrial level.

The next sections will discuss the main alternatives to replace HF as a catalyst in industrial processes to maintain optimal yields in the alkylate reaction and avoid the challenges of its use in alkylation units.

5.3   OTHER CATALYSTS USED IN THE ALKYLATION REACTION

The following sections describe the problems, challenges, safety considerations, disposal of wastes, and expensive devices available for handling corrosive acids such as H_2SO_4 and HF. Additionally, other inconveniences of using HF include its high volatility and rapid dispersion over several kilometers and the high operation cost due to the use of corrosion-resistant materials, which have resulted in the identification of new alternatives for the use of other catalysts capable of maintaining the quality of alkylates in industrial processes.[50]

This section discusses recent advances in finding technological solutions to replace H_2SO_4 and HF with promising alkylation catalysts that improve the reactivity and selectivity of alkylation products for gasoline blending. These new materials are catalysts based on Brönsted acids (acidic chloroaluminates), ionic liquids (ILs), zeolites, or their combinations.

5.3.1 IONIC LIQUIDS (ILs)

From the challenge discussed above in using H_2SO_4 and HF as catalysts, ILs have emerged as alternatives or promoters of alkylation reactions. ILs have unique and interesting physicochemical properties such as being liquid salts at room temperature, acting as polar solvents and ionic salts, nonvolatility, thermal stability, ionic conductivity, and recyclability, and these properties are modulable to enhance or obtain a great variety of these compounds. In this way, ILs have been found in various applications, such as hydrogenation, oxidation, extraction, and isomerization. Moreover, catalytic applications for obtaining high-quality gasoline alkylates have been explored.[53]

If an IL is used, it needs a series of characteristics to be considered a good catalyst in an alkylation reaction: high acidity, such as a Lewis or Brönsted acid. These properties could be modulated through the design of the anionic or cationic fragments; in this sense, a recent review has provided a complete discussion about the use of ILs in alkylation reactions.[54] The mechanism of the alkylation reaction is similar to that using mineral acids, with differences in the type of acidity provided by the IL.

For example, Brönsted acidic ILs are based on $[N-H]^+$-type protic cations that work as proton donors and the anion stabilizes the species during the catalytic cycle. On the other hand, other ILs are based on sulfonic groups in the anion fragment and act as proton donors. In contrast, ILs based on Lewis acids are also used. The main reason is that Brönsted acidity is insufficient to achieve the best conversion or selectivity. For Lewis acid-based ILs, the presence of strong Lewis acids such as $Al_2O_7^-$, $Zn_2Cl_5^-$, $CuAlCl_5^-$ etc. in the anion fragment increases the acidity of the IL. Combinations of Brönsted–Lewis acids in the structure of ILs also exist to establish a synergistic effect to increase the acidity of the whole compound. There are other combinations where the acidic IL is mixed with mineral acids, such as HCl or mixed with inorganic salts, such as Lewis acids, for example, CuCl, $AlCl_3$, $FeCl_3$, $ZnCl_5$, $CuCl_2$, AgCl, $ZnCl_2$, $NiCl_2$, $CeCl_3$, $SnCl_2$, ZnO, and CuO.[54]

The general structures of ions featured in ILs with Brönsted acidity, Lewis acidity, and their synergistic combinations used as catalysts in alkylation reactions are shown below in Figure 5.7.

The mechanism of the alkylation reaction using an IL is similar to that of the conventional catalytic cycle with mineral acids (Figure 5.8): once the IL reacts with the olefin to form the carbenium species, it reacts with another alkene to undergo the alkylation process, followed by an isomerization step, and finally, hydride transfer to produce the alkylate product, recovering another carbenium ion to continue the catalytic cycle.[39]

Additionally, organic or inorganic additives are sometimes added to increase the catalytic activity of ILs. For example, if CuCl, $CuCl_2$, or $NiCl_2$ is incorporated into

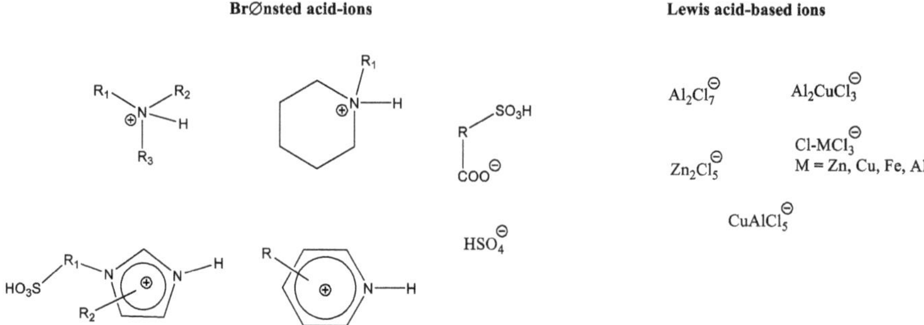

FIGURE 5.7 General structures of ions in ILs based on Brönsted and Lewis acids.

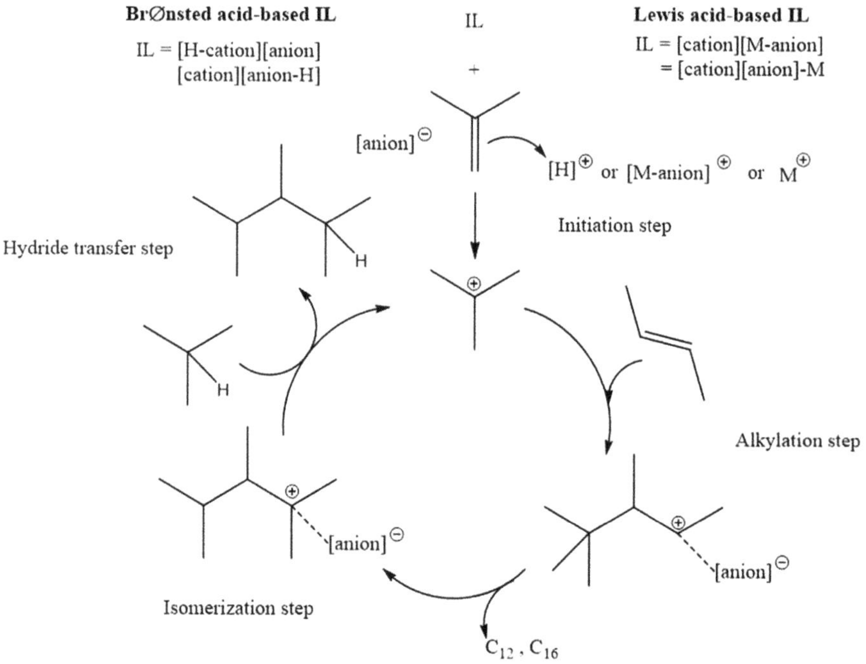

FIGURE 5.8 Mechanism of the alkylation reaction using ILs with Brönsted and Lewis acid substituents.

an IL, the alkylate quality could be improved. Complementarily, butyl-thioalcohol, ethyl-thioether, benzene-chloroaluminate, toluene, xylenes, alkyl-benzenes, and tri-ethylamine hydrochloride additives have been incorporated. These additives generally improve the alkylate yield and selectivity to TMP and the RON.[54]

In addition, ILs have also been supported on zeolites. The study of the immobilization of acidic ILs on mesoporous zeolites has been explored, and these new catalysts have been shown to enhance alkylate reactions. The main zeolites or supports used to immobilize ILs are MCM-41, ZS-MCM-41, ZS, Y, and SBA, among others.

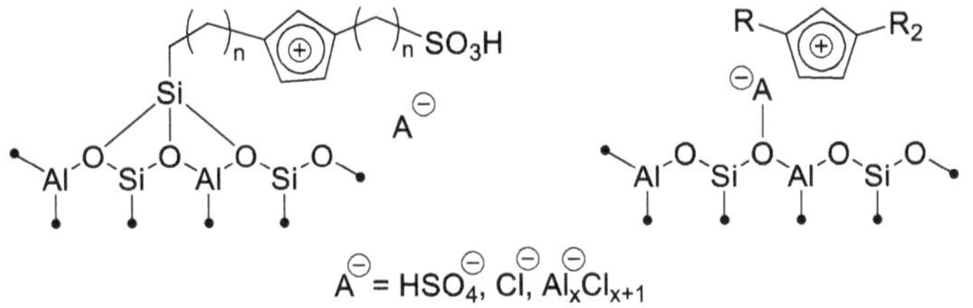

FIGURE 5.9 Schematic structure of ILs immobilized on a zeolite surface.

Of course, the employed ILs need to have the characteristics of Brönsted or Lewis acid groups, which in turn, contribute to the acidity of the support and increase both the alkylate yield and selectivity to TMP. Some examples of used ILs are based on imidazolium cations, such as 1-butyl-3 triethoxysilylpropyl-imidazolium hydrogen sulfate.[54] However, these new catalysts are currently under investigation. Figure 5.9 shows a general scheme of the support and IL immobilization.

It is important to highlight that ILs can be prepared at the industrial level;[55] this fact represents an opportunity to implement them in industrial processes.[56] In this sense, there are two examples of the use of ILs in alkylation plant refineries. The first one is the Isoalky™ technology[57] developed by Chevron and Honeywell UOP and installed in Salt Lake City. This refinery is the first in the world to implement an alkylation unit based on an IL capable of producing 5000 barrels/day of alkylate.[58] This process represents a major innovation in alkylation technology. The second one is the Ionikilatyon™ technology[59] developed by the China University of Petroleum and licensed by Well Resources from Canada; now, six units in China have been installed to produce 7500 barrels/day of alkylate. These two industrial technologies are described below.

5.3.1.1 Isoalky™ Technology

The catalyst in the Isoalky™ process is a chloroaluminate-based IL, where the cation is an ammonium or phosphonium ion and the anion is $AlCl_4^-$ or $Al_2Cl_7^-$. This catalyst exhibits high activity and selectivity to alkylation and long thermal stability. Additionally, a trace amount of HCl is needed as a cocatalyst, which is generated *in situ* through the addition of an organic chloride promoter.[60] This technology exhibits a series of benefits:

- The refinery produces an equal or greater volume of alkylate with less catalyst amount and a smaller reactor (3–6% of catalyst).
- The reaction produces higher octane gasoline or maintains the conventional gasoline pool.
- The Isoalky™ unit provides the refinery with the option to use whichever feedstock.
- Isoalky™ technology could upgrade C3–C5 with maximum feed efficiency.
- There is a low degree of polymerization of olefins. In other words, the alkylate yield is higher. On the other hand, polymers are converted into regenerated naphtha and blended back into alkylate gasoline, providing additional yield.

- Reduced environmental impact due to the fact that online regeneration eliminates polymer or catalyst incineration. A smaller reactor generates a smaller volume or reduces the amount of acidic and caustic products.

5.3.1.2 Ionikilatyon™ Technology

PetroChina Dagang Petrochemical Co. has commissioned a new unit based on IL alkylation technology to produce high-octane alkylate in its 5 million/year refinery.

The Ionikilatyon™ technology is based on nontoxic ILs with a fraction of $AlCl_3$, HCl, and $CuCl_2$. This IL-based composite is the catalyst and has adequate handling behavior. The process consists of several stages: the first one is feed pretreatment and the second one is the alkylation reaction, followed by the separation of products and catalyst regeneration.[61–63] Ionikilatyon™ offers the following advantages over conventional alternatives:

- Innovative technology with multiple commercial units.
- The use of corrosive HF and H_2SO_4 catalysts was eliminated.
- Safer process, noncorrosive system, and low-cost carbon steel equipment.
- Nonhazardous emissions of waste and reduced energy consumption: environmentally friendly.
- High catalyst activity and stable reactor operation conditions.
- High-quality alkylate products (96–98 RON).
- Integrated catalyst regeneration.

This fact can be seen after the analysis of two main industrial technologies. The beneficial aspects are excellent, and they are significant in the safer operation, adaptability of the technologies over the existing refineries, and, of course, the successful commercial test.

5.3.2 Zeolites

It is well known that heterogeneous catalysts are widely used in crude oil refining catalytic processes, such as FCC, Fisher–Tropsch, hydrotreatment processes, etc. The following section discusses the use of Brönsted and Lewis acids such as $AlCl_3$ SbF_5, or BF_3-based catalysts. However, other inorganic solids that exhibit acidic properties, such as zeolites, zirconia, a mixture of oxides, or acidic resin, have been studied by many researchers for alkylation reactions.[26,50]

In this sense, zeolites have a large surface area due to their large porosity, exhibit acidic properties, have exceptional thermal stability, and have the ability to undergo hydride transfer. However, rapid deactivation due to coking represents an operational disadvantage because zeolites depend on additional regeneration processes. Although this operation is well known in the oil industry, it does not represent a serious challenge. Among the zeolites that are more commonly employed in alkylation reactions are Y-zeolites, X-zeolites, MCM-22, USY, ZSM, and SBA; some of these incorporate promoters or cocatalyst substances to enhance or improve the acidic properties, such as Pt, Zr, Ce, Cu, Zn, Al, or HCl or CF_3SO_3H acids, and as it was mentioned above, functionalized ILs (Figure 5.10).[50,54,64]

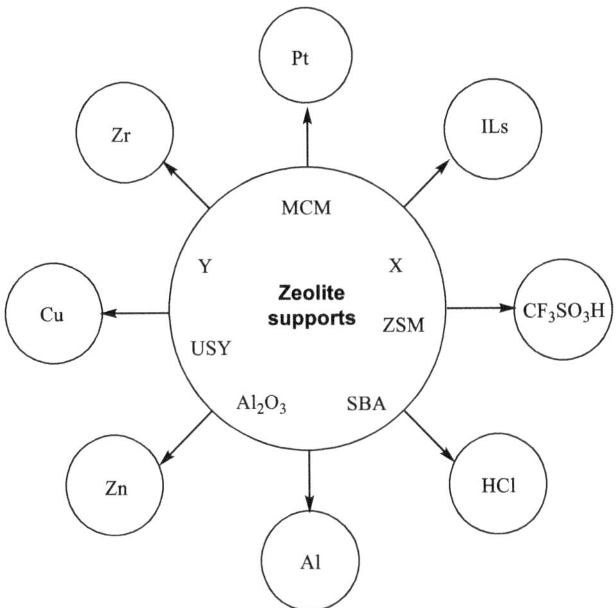

FIGURE 5.10 Main zeolite supports and promoters used in the alkylation process.

Of course, zeolite-based catalysts for the alkylation process have been developed commercially. Here are some examples:

5.3.2.1 AlkyClean™ Process

This technology was developed by CB&I, Albemarle Corp., and Neste Corp. It was the first commercial-scale process based on a solid acid catalyst. The unit for alkylation has a capacity of 2700 barrels/day for alkylate production. The technology employs Pt/US-Y, a fixed zeolite developed by Albemarle with the trade name AlkyStar. The general process uses a series of reactors with distributed injection olefin feeds to achieve high isobutane/alkene ratios. This process requires an additional regeneration step at 250°C in a hydrogen atmosphere.[50]

5.3.2.2 K-SAAT™ Technology

The next-generation solid acid alkylation technology was developed by M/s KBR and Exelus. The catalyst is a zeolite Y-based ExSact and a precious metal-free material, which provides superior alkylation performance with an alkylate with RON >99%. This K-SAAT process is more efficient, cost-effective, and environmentally benign than liquid mineral acids. KBR has signed two K-SAAT licenses: the first one to Dongying Haike Ruilin Chemical Co. Ltd., in Shandong Province, China, and the second one to Luoyang Aiyou Chemical Co., in Henan Province, China.[65]

5.3.2.3 Lurgi EuroFuel® and Haldor Topsoe FBA™ Processes

Lurgi and Süid-Chemie AG developed another zeolite-based technology for the alkylation process known as EuroFuel®, consisting of a Y-zeolite with a high

TABLE 5.2

Comparison of the Different Technologies Used for the Alkylation Process in the Industry

Catalyst	T (°C)	P (psi)	Isobutane/ Olefin Ratio	RON	Catalyst %	Corrosion Level
H_2SO_4	0–15	30–80	7–15	95–96	50	High
HF	35	200	10	95	60–75	High
Isoalky™	0–50	40–250	8	94–99	3–6	Low
Ionikylation™	10–25	110–170	8–10	96–98	ND	Low
AlkyClean™	50–90	1–100	8–10	91–94	20–80	Low
EuroFuel®	50–100	ND	6–12	ND	ND	Low
K-SAAT™	60–71	ND	8–15	>99	ND	Low

ND: not available

concentration of Brönsted acid sites and lower Lewis acid sites. The reactor is based on a distillation tower concept, where the catalyst is mixed with the isobutane feed and the alkene at the top isobutane mixture is introduced at stages: the catalyst is mixed as a consequence of boiling the mixture of alkylate and isobutane. Haldor Topsoe has developed a fixed-based alkylation (FBA™) process that employs a liquid catalyst: trifluoromethanesulfonic acid supported on a zeolite.[66] Information about these processes is scarce due to them being maintained as an industrial secret.

Table 5.2 shows the main characteristics of the alkylation processes using mineral acids (conventional catalysts), IL technologies, and zeolite-based technologies implemented at the industrial level in the alkylation process to obtain gasoline.

5.4   CONCLUSIONS

This chapter analyzed and compared conventional alkylation processes, which use H_2SO_4 and HF, and are now firmly established industrial processes for obtaining almost all alkylation gasoline globally. New technologies based on solid catalysts, such as the AlkyClean process, K-SAAT technology, Lurgi EuroFuel, and Haldor Topsoe FBA processes, and ILs, such as the Isoalky technology and the Ionikylation technology, were also discussed. Technologies using ILs as catalysts have proven to be a viable alternative to those using strong acids such as H_2SO_4 and HF because they have high conversion and selectivity, the catalysts can be regenerated, and they are considered more environmentally friendly.

REFERENCES

(1) *The number of cars worldwide is set to double by 2040*. World Economic Forum. https://www.weforum.org/agenda/2016/04/the-number-of-cars-worldwide-is-set-to-double-by-2040/ (accessed January 11, 2024).

(2) *World Energy Scenarios: Global Transport Scenarios 2050*. World Energy Council. https://www.worldenergy.org/publications/entry/world-energy-scenarios-global-transport-scenarios-2050 (accessed January 11, 2024).

(3) Change, I. C. Mitigation of Climate Change. Contrib. Work. Group III Fifth Assess. Rep. Intergov. Panel Clim. Change 2014, *1454*, 147.

(4) Organization of the Petroleum Exporting Countries. 2013 World Oil Outlook, 2013.

(5) Kalghatgi, G. Is It Really the End of Internal Combustion Engines and Petroleum in Transport? *Appl. Energy* 2018, *225*, 965–974. https://doi.org/10.1016/j.apenergy.2018.05.076

(6) Sarathy, S. M.; Farooq, A.; Kalghatgi, G. T. Recent Progress in Gasoline Surrogate Fuels. *Prog. Energy Combust. Sci.* 2018, *65*, 67–108. https://doi.org/10.1016/j.pecs.2017.09.004

(7) Aklilu, A. Z. Gasoline and Diesel Demand in the EU: Implications for the 2030 Emission Goal. *Renew. Sustain. Energy Rev.* 2020, *118*, 109530. https://doi.org/10.1016/j.rser.2019.109530

(8) Abdellatief, T. M. M.; Ershov, M. A.; Savelenko, V. D.; Kapustin, V. M.; Makhova, U. A.; Klimov, N. A.; Chernysheva, E. A.; Aboul-Fotouh, T. M.; Abdelkareem, M. A.; Mustafa, A.; Olabi, A. G. Advanced Progress and Prospects for Producing High-Octane Gasoline Fuel toward Market Development: State-of-the-Art and Outlook. *Energy Fuels* 2023, *37* (23), 18266–18290. https://doi.org/10.1021/acs.energyfuels.3c02541

(9) Ivanchina, E. D.; Kirgina, M. V.; Chekantsev, N. V.; Sakhnevich, B. V.; Sviridova, E. V.; Romanovskiy, R. V. Complex Modeling System for Optimization of Compounding Process in Gasoline Pool to Produce High-Octane Finished Gasoline Fuel. *Chem. Eng. J.* 2015, *282*, 194–205. https://doi.org/10.1016/j.cej.2015.03.014

(10) Hsu, C. S.; Robinson, P. R. Gasoline Production and Blending. In *Springer Handbook of Petroleum Technology*; Hsu, C. S., Robinson, P. R., Eds.; Springer Handbooks; Springer International Publishing: Cham, 2017; pp 551–587. https://doi.org/10.1007/978-3-319-49347-3_17

(11) Ershov, M. A.; Grigorieva, E. V.; Abdellatief, T. M. M.; Kapustin, V. M.; Abdelkareem, M. A.; Kamil, M.; Olabi, A. G. Hybrid Low-Carbon High-Octane Oxygenated Gasoline Based on Low-Octane Hydrocarbon Fractions. *Sci. Total Environ.* 2021, *756*, 142715. https://doi.org/10.1016/j.scitotenv.2020.142715

(12) Abdellatief, T. M. M.; Ershov, M. A.; Kapustin, V. M.; Ali Abdelkareem, M.; Kamil, M.; Olabi, A. G. Recent Trends for Introducing Promising Fuel Components to Enhance the Anti-Knock Quality of Gasoline: A Systematic Review. *Fuel* 2021, *291*, 120112. https://doi.org/10.1016/j.fuel.2020.120112

(13) Kuchling, T.; Endisch, M.; Schneider, J.; Kureti, S.; Hübner, W.; Preis, M.; Schmidt, C. Potential of On-Board Gasoline Upgrading for Enhancement of Engine Efficiency. *Chem. Eng. Technol.* 2017, *40* (9), 1644–1651. https://doi.org/10.1002/ceat.201600464

(14) Hancsók, J. Gasoline Fuels for Spark-Ignition Internal Combustion Engines. In *Kirk-Othmer Encyclopedia of Chemical Technology*; John Wiley & Sons, Ltd, 2016; pp 1–60. https://doi.org/10.1002/0471238961.0701191508150308.a01.pub3

(15) *Corporate Average Fuel Economy|NHTSA.* https://www.nhtsa.gov/laws-regulations/corporate-average-fuel-economy (accessed January 11, 2024).

(16) Ghosh, P.; Hickey, K. J.; Jaffe, S. B. Development of a Detailed Gasoline Composition-Based Octane Model. *Ind. Eng. Chem. Res.* 2006, *45* (1), 337–345. https://doi.org/10.1021/ie050811h

(17) Taylor, C. F. *Internal Combustion Engine in Theory and Practice, Second Edition, Revised, Volume 2: Combustion, Fuels, Materials, Design*; MIT Press, 1985.

(18) Anderson, J. E.; Wallington, T. J.; Stein, R. A.; Studzinski, W. M. Issues with T50 and T90 as Match Criteria for Ethanol-Gasoline Blends. *SAE Int. J. Fuels Lubr.* 2014, *7* (3), 1027–1040.

(19) Ershov, M. A.; Savelenko, V. D.; Shvedova, N. S.; Kapustin, V. M.; Abdellatief, T. M. M.; Karpov, N. V.; Dutlov, E. V.; Borisanov, D. V. An Evolving Research Agenda of Merit Function Calculations for New Gasoline Compositions. *Fuel* 2022, *322*, 124209. https://doi.org/10.1016/j.fuel.2022.124209

(20) Duan, X.; Feng, L.; Liu, H.; Jiang, P.; Chen, C.; Sun, Z. Experimental Investigation on Exhaust Emissions of a Heavy-Duty Vehicle Powered by a Methanol-Fuelled Spark Ignition Engine under World Harmonized Transient Cycle and Actual on-Road Driving Conditions. *Energy* 2023, *282*, 128869. https://doi.org/10.1016/j.energy.2023.128869

(21) American Fuel & Petrochemical Manufacturers. *Alkylate: Understanding a Key Component of Cleaner Gasoline.* https://www.afpm.org/newsroom/blog/alkylate-understanding-key-component-cleaner-gasoline (accessed January 17, 2024).

(22) Vora, B. V.; Kocal, J. A.; Barger, P. T.; Schmidt, R. J.; Johnson, J. A. Alkylation. *Kirk-Othmer Encyclopedia of Chemical Technology*; Wiley-Interscience, 2007; Vol. 2, pp 169–203.

(23) Ipatieff, V. N.; Grosse, A. V. Reaction of Paraffins with Olefins. *J. Am. Chem. Soc.* 1935, *57* (9), 1616–1621. https://doi.org/10.1021/ja01312a034

(24) Ipatieff, V. N. *Catalytic Reactions at High Pressures and Temperatures*; Macmillan, 1936.

(25) Albright, L. F. Alkylation—Industrial. In *Encyclopedia of Catalysis*; John Wiley & Sons, Ltd, 2010. https://doi.org/10.1002/0471227617.eoc014.pub2

(26) Schmerling, L., Boord, C. E., Kurtz, S. S., Brooks, B. T., Eds.; *The Chemistry of Petroleum Hydrocarbons*; Reinhold Publishing Corporation, 1954.

(27) Himes, J. F.; Mehlberg, R. L.; Pujadó, P. R.; Ward, D. J. Gasoline Components. In *Handbook of Petroleum Processing*; Jones, D. S. J. S., Pujadó, P. R., Eds.; Springer Netherlands: Dordrecht, 2006; pp 355–416. https://doi.org/10.1007/1-4020-2820-2_9

(28) Akhmadova, Kh Kh; Magomadova, M. Kh; Syrkin, A. M.; Egutkin, N. L. History, Current State, and Prospects for Development of Isobutane Alkylation with Olefins. *Theor. Found. Chem. Eng.* 2019, *53* (4), 643–655. https://doi.org/10.1134/S0040579519040092

(29) Feller, A.; Zuazo, I.; Guzman, A.; Barth, J. O.; Lercher, J. A. Common Mechanistic Aspects of Liquid and Solid Acid Catalyzed Alkylation of Isobutane with N-Butene. *J. Catal.* 2003, *216* (1), 313–323. https://doi.org/10.1016/S0021-9517(02)00068-4

(30) Hofmann, J. E.; Schriesheim, A. Ionic Reactions Occurring During Sulfuric Acid Catalyzed Alkylation. I. Alkylation of Isobutane with Butenes. *J. Am. Chem. Soc.* 1962, *84* (6), 953–957. https://doi.org/10.1021/ja00865a015

(31) Esteves, P. M.; Araújo, C. L.; Horta, B. A. C.; Alvarez, L. J.; Zicovich-Wilson, C. M.; Ramírez-Solís, A. The Isobutylene–Isobutane Alkylation Process in Liquid HF Revisited. *J. Phys. Chem. B* 2005, *109* (26), 12946–12955. https://doi.org/10.1021/jp051567a

(32) Albright, L. F. Alkylation of Isobutane with C3–C5 Olefins: Feedstock Consumption, Acid Usage, and Alkylate Quality for Different Processes. *Ind. Eng. Chem. Res.* 2002, *41* (23), 5627–5631. https://doi.org/10.1021/ie020323z

(33) Ivashkina, E.; Ivanchina, E.; Dolganov, I.; Chuzlov, V.; Kotelnikov, A.; Dolganova, I.; Khakimov, R. Nonsteady-State Mathematical Modelling of H_2SO_4-Catalysed Alkylation of Isobutane with Alkenes. *Oil Gas Sci. Technol. – Rev. D'IFP Energ. Nouv.* 2021, *76*, 36. https://doi.org/10.2516/ogst/2021017

(34) Cao, P.; Zheng, L.; Sun, W.; Zhao, L. Multiscale Modeling of Isobutane Alkylation with Mixed C4 Olefins Using Sulfuric Acid as Catalyst. *Ind. Eng. Chem. Res.* 2019, *58* (16), 6340–6349. https://doi.org/10.1021/acs.iecr.9b00874

(35) Corma, A.; Martínez, A. Chemistry, Catalysts, and Processes for Isoparaffin–Olefin Alkylation: Actual Situation and Future Trends. *Catal. Rev.* 1993, *35* (4), 483–570. https://doi.org/10.1080/01614949308013916

(36) Li, K. W.; Eckert, R. E.; Albright, L. F. Alkylation of Isobutane with Light Olefins Using Sulfuric Acid. Operating Variables Affecting Physical Phenomena Only. *Ind. Eng. Chem. Process Des. Dev.* 1970, *9* (3), 434–440. https://doi.org/10.1021/i260035a011

(37) Alkylation and Polymerization; Meyers, R. A., Ed.; McGraw-Hill Education: New York, 2016.

(38) Albright, L. F. Improving Alkylate Gasoline Technology. *Improv. Alkylate Gasol. Technol.* 1998, *28* (7), 46–53.

(39) Zhang, H.; Liu, R.; Yang, Z.; Huo, F.; Zhang, R.; Li, Z.; Zhang, S.; Wang, Y. Alkylation of Isobutane/Butene Promoted by Fluoride-Containing Ionic Liquids. *Fuel* 2018, *211*, 233–240. https://doi.org/10.1016/j.fuel.2017.09.029

(40) Xin, Y.; Hu, Y.; Li, M.; Chi, K.; Zhang, S.; Gao, F.; Jiang, S.; Wang, Y.; Ren, C.; Li, G. Isobutane Alkylation Catalyzed by H_2SO_4: Effect of H_2SO_4 Acid Impurities on Alkylate Distribution. *Energy Fuels* 2021, *35* (2), 1664–1676. https://doi.org/10.1021/acs.energyfuels.0c03453

(41) Albright, L. F.; Wood, K. V. Alkylation of Isobutane with C3–C4 Olefins: Identification and Chemistry of Heavy-End Production. *Ind. Eng. Chem. Res.* 1997, *36* (6), 2110–2120. https://doi.org/10.1021/ie960265f

(42) Sun, W.; Shi, Y.; Chen, J.; Xi, Z.; Zhao, L. Alkylation Kinetics of Isobutane by C4 Olefins Using Sulfuric Acid as Catalyst. *Ind. Eng. Chem. Res.* 2013, *52* (44), 15262–15269. https://doi.org/10.1021/ie400415p

(43) Liu, Z.; Meng, X.; Zhang, R.; Xu, C.; Dong, H.; Hu, Y. Reaction Performance of Isobutane Alkylation Catalyzed by a Composite Ionic Liquid at a Short Contact Time. *AIChE J.* 2014, *60* (6), 2244–2253. https://doi.org/10.1002/aic.14394

(44) Boronat, M.; Viruela, P.; Corma, A. Theoretical Study of the Mechanism of Branching Rearrangement of Carbenium Ions. *Appl. Catal. Gen.* 1996, *146* (1), 207–223. https://doi.org/10.1016/0926-860X(96)00160-3

(45) Liu, Y.; Wang, L.; Li, R.; Hu, R. Reaction Mechanism of Ionic Liquid Catalyzed Alkylation: Alkylation of 2-Butene with Deuterated Isobutene. *J. Mol. Catal. Chem.* 2016, *421*, 29–36. https://doi.org/10.1016/j.molcata.2016.05.005

(46) Bui, T. L. T.; Korth, W.; Aschauer, S.; Jess, A. Alkylation of Isobutane with 2-Butene Using Ionic Liquids as Catalyst. *Green Chem.* 2009, *11* (12), 1961–1967. https://doi.org/10.1039/B913872B

(47) Lee, L.; Harriott, Peter. The Kinetics of Isobutane Alkylation in Sulfuric Acid. *Ind. Eng. Chem. Process Des. Dev.* 1977, *16* (3), 282–287. https://doi.org/10.1021/i260063a006

(48) Albright, L. F.; Spalding, M. A.; Faunce, J.; Eckert, R. E. Alkylation of Isobutane with C4 Olefins. 3. Two-Step Process Using Sulfuric Acid as Catalyst. *Ind. Eng. Chem. Res.* 1988, *27* (3), 391–397. https://doi.org/10.1021/ie00075a005

(49) Liang, L.; Liu, Y.; Jiao, W.; Zhang, Q.; Zhang, C. Octane Compositions in Sulfuric Acid Catalyzed Isobutane/Butene Alkylation Products: Experimental and Quantum Chemistry Studies. *Front. Chem. Sci. Eng.* 2021, *15* (5), 1229–1242. https://doi.org/10.1007/s11705-020-2030-x

(50) Pant, K. K., Gupta, S. K., Ahmad, E., Eds.; *Catalysis for Clean Energy and Environmental Sustainability: Petrochemicals and Refining Processes - Volume 2*, Springer International Publishing: Cham, 2021. https://doi.org/10.1007/978-3-030-65021-6

(51) Hussain, N. *Alkylation Process In Petroleum Refinery.* https://thepetrosolutions.com/alkylation-process-petroleum-refiney/

(52) Simpson, M. B.; Kester, M. Hydrofluoric Acid Alkylation. *ABB Rev.* 2007, 22–26.

(53) Rogers, R. D., Seddon, K. R., Eds.; *Ionic Liquids: Industrial Applications for Green Chemistry*, ACS Symposium Series; American Chemical Society: Washington, DC, 2002; Vol. 818. https://doi.org/10.1021/bk-2002-0818

(54) Guzmán-Lucero, D.; Guzmán-Pantoja, J.; Velázquez, H. D.; Likhanova, N. V.; Bazaldua-Domínguez, M.; Vega-Paz, A.; Martínez-Palou, R. Isobutane/Butene Alkylation Reaction Using Ionic Liquids as Catalysts. Toward a Sustainable Industry. *Mol. Catal.* 2021, *515*, 111892. https://doi.org/10.1016/j.mcat.2021.111892

(55) Short, P. OUT OF THE IVORY TOWER Ionic Liquids Are Starting to Leave Academic Labs and Find Their Way into a Wide Variety of Industrial Applications. *Chem. Eng. News* 2006.

(56) Tullo, A. H. Oil Company Begins Major New Use of Ionic Liquids. *Chemical & Engineering News*. April 16, 2021. https://cen.acs.org/materials/ionic-liquids/Oil-company-begins-major-new/99/web/2021/04 (accessed February 21, 2024).

(57) Chevron. *Chevron and Honeywell Announce Start-up of World's First Commercial ISOALKY™ Ionic Liquids Alkylation Unit.* chevron.com. https://www.chevron.com/stories/chevron-and-honeywell-announce-start-up-of-isoalky-ionic-liquids-alkylation-unit (accessed May 05, 2021).

(58) Timken, H.-K.; Gattupalli, R.; Tertel, J. A.; Weber, D. *A New Era for Alkylation: Ionic Liquid Alkylation* - Isoalky Process Technology, 2020.

(59) *Ionikylation|Well Resources*. Well Resources Inc. https://www.wellresources.ca/ionikylation (accessed February 21, 2024).

(60) *ISOALKYTM Technology: Next-Generation Alkylate Gasoline Manufacturing Process Technology Using Ionic Liquid Catalyst.* springerprofessional.de. https://www.springer professional.de/en/isoalky-technology-next-generation-alkylate-gasoline-manufacturi/17702208 (accessed February 27, 2024).

(61) Brelsford, R. PetroChina refinery starts up Ionikylation unit. *Oil & Gas Journal*. https://www.ogj.com/refining-processing/refining/article/14281885/petrochina-refinery-starts-up-ionikylation-unit (accessed February 21, 2024).

(62) *Second Ionikylation Unit Commissioned*. Well Resources Inc. https://www.wellresources.ca/ionikylation-commissioned-china (accessed February 21, 2024).

(63) Warren, C. *Second Brownfield Composite Ionic Liquid Alkylation Unit Commissioned in China.* Ionikylation. https://www.wellresources.ca/ionikylation-commissioned-china (accessed April 25, 2021).

(64) Feller, A. Reaction Mechanism and Deactivation Pathways in Zeolite Catalyzed Isobutane/2-Butene Alkylation.

(65) *KBR Technology Continues to Expand Portfolio|KBR*. https://www.kbr.com/en/insights-news/stories/kbr-technology-continues-expand-portfolio (accessed February 27, 2024).

(66) Sagaard-Andersen, P.; Hommeltoft, S. I.; Sarup, B. *[9]P1 Fixed-Bed Alkylation –Next Generation Alkylation Process*; OnePetro, 1997.

6 Gasoline Octane Enhancers

Oxygenated and Non-oxygenated Additives

Luis Felipe Cruz Martínez
Universidad Autónoma Metropolitana-Azcapotzalco,
Mexico City, México

Javier Guzmán-Pantoja
Mexican Petroleum Institute, Mexico City, México

6.1 INTRODUCTION

Throughout history, there have been countless theoretical contributions, inventions, modifications, and experiments that have become fundamental to the development of the internal combustion engine. In the early 19th century, the quest for a more efficient means of transportation than that powered by equine and cattle led to the invention of the steam engine, which transformed heat into mechanical energy by means of an external combustion process. Notwithstanding, this type of engine was inefficient and impractical, so research into engine development continued. In 1824, a French mechanical engineer Sadi Carnot established the basics of the processes and theoretical conditions for the correct functioning of internal combustion engines. Étienne Lenoir, in 1860, created a two-stroke internal combustion engine vehicle powered by a mixture of air and coal gas. Although not very efficient, it was established as the power source for the first vehicle that was not powered by animals or a steam engine and set the tone for the development and improvement of explosive combustion systems. In 1862, Alphonse Beau de Rochas conducted research on the thermodynamic principles of the processes of a gas in a rigid cylinder capable of supporting the operation of a four-stroke internal combustion engine. In 1867, Nikolaus Otto developed an engine that was slightly more efficient than Lenoir's. Otto continued conducting research in the field, and in 1876, the first four-stroke internal combustion engine, also called the Otto cycle engine, was created, which used a mixture of air and gasoline.[1]

The Otto cycle engine has been the most widely used engine in motor vehicles to date because of its economy and performance. This type of engine performs a

DOI: 10.1201/9781003517283-6

"

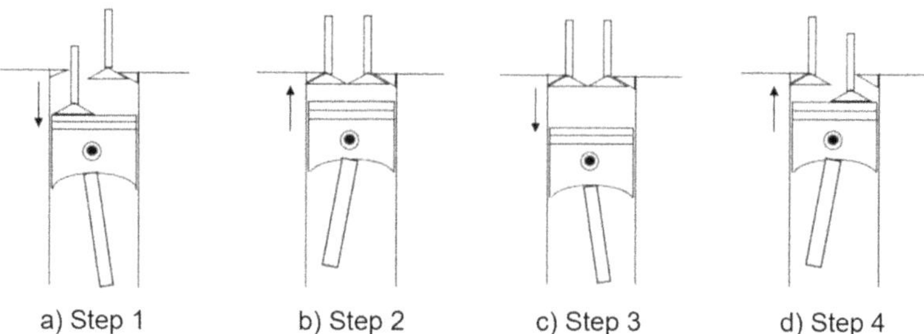

FIGURE 6.1 Description of the stages of an internal combustion engine. (a) Admission, (b) compression, (c) detonation, and (d) expulsion.

thermodynamic combustion cycle in four stages. At the initial stage, fuel/comburent (a mixture of gasoline and air) is admitted to the combustion chamber by opening the valves. Once the mixture has entered, the valves are closed, and the gases contained in the chamber are compressed. At this point, the third stage occurs: the detonation of the fuel–air mixture with the aid of a spark. Finally, the generated gases and remaining hydrocarbons (HCs) are expelled, thus repeating the four-stage cycle.[2] The process is represented in Figure 6.1.

Each stage of this engine is fundamental, as an adequate fuel–air mixture ratio is required, as well as good fuel properties to achieve the highest compression and to avoid premature explosion problems that cause irreparable damage to the internal combustion engine.

With the modernization of fuels, various modifications to the internal combustion engines have been implemented; however, these modifications have not been sufficient to achieve optimum performance. Therefore, several measurement techniques are used to evaluate fuel properties that delay detonation, improve the co-firing of gases, and increase engine lifetime. In this context, the most measured property to determine these advantages in fuels is the one known as octane number.

6.2 OCTANE RATING

Octane rating, also known as octane number, measures the ability of fuels to be compressed before detonation occurs. Two values are used to measure octane rating: the research octane number (RON) and motor octane number (MON). These tests are based on the analysis of n-heptane and 2,2,4-trimethylpentane (isooctane) mixtures, where their antiknock properties are determined as a function of the compression ratio. The octane rating of n-heptane is 0, while that of isooctane is 100. The mixture of these two components is called the primary reference fuel (PRF) and is measured on a cooperative fuel research (CFR) engine.

The RON and MON values are quantified with the standardized methods ASTM D2699-23b and ASTM D2700-23b, respectively. The RON test represents the combustion quality under controlled conditions, performed on a single-cylinder engine at an intake temperature of 52°C and 600 rpm, while the MON test represents the quality of gasoline under normal operating conditions (intake temperature greater than 150°C and 900 rpm).

Several modeling studies proposing novel experiments to determine the RON and MON values in fuels have been reported. The CFR method, implemented since the 1930s, has become inaccurate with the use of more complex modern fuels.[3,4] Additionally, another parameter called octane sensitivity (OS), which is the difference between the RON and MON values, has been proposed. This sensitivity is influenced by the chemical structure of various compounds in the fuel, which affect the reactions produced during combustion. The OS level is highly variable, as the addition of compounds in modern fuels has become increasingly common.[5] The modification of PRFs by adding toluene (TPRF) has also been proposed, as PRFs are not representative samples of modern fuels to give an approximation of the octane number. There are variations of up to 10 units in the RON or MON values.[6]

The regularly measured value for determining the antiknock capability of fuels is the RON. A high RON value is indicative of a high-octane rating. However, a high-octane number slows down the flame and causes a delay in reaching the maximum combustion pressure. In a study by Athafah and Salih,[7] it was observed that using a fuel with a higher octane rating than that required by the engine did not substantially increase the effective power output, but did increase fuel consumption. In another study by Sayin et al.,[8] it was found that using a fuel with a higher octane rating than the one required by the engine increased carbon monoxide (CO) emissions by up to 5%.

In the case of gasoline, a low-octane rating gives it a lower flash point, which considerably reduces combustion time, increases CO, HC, and nitrogen oxide (NO_x) emissions, and causes premature explosions that lead to harmful knocking in the engine. The knocking that occurs in the engine results from the self-ignition of the fuel–air mixture contained in the combustion chamber before it can be consumed by the flame.[4]

Due to the generated knocking, which causes irreparable damage to the engine over a long period of use, and considering that the efficiency of the engine will always depend on the type of fuel used, it has become essential to regulate the octane number. Gasoline is a fuel composed of hundreds of HCs, mainly alkanes, paraffins, olefins, aromatics, and naphthenes. The composition of gasoline varies according to its origin, depending on the crude oil and the proportion of the streams that are mixed to form it, which in turn are obtained from various refining processes.

The octane number is not the same for all petrol types. Figure 6.2 contains data from a study conducted by Abdellatief.[9] These data are used only as a representative sample. However, although variable RON levels are observed depending on the source of the fuel stream, the most frequently used and marketed gasoline is a blend of HCs and additives, which taken together have adequate antiknock capability for internal combustion engines. Failure to properly formulate the gasoline blend would result in premature explosions, higher pollutant emissions from unburned gases, and low engine power. It is therefore important to add various compounds to increase the octane rating.

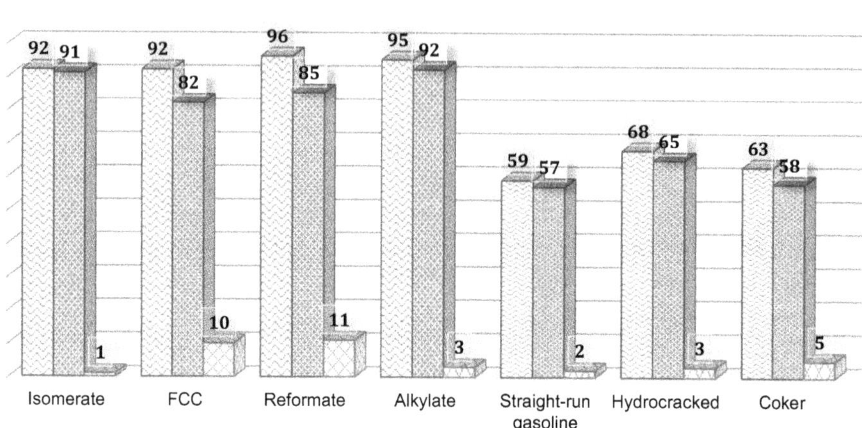

FIGURE 6.2 RON, MON, and OS values for the different types of petrol. From data reported by Abdellatief.[9]

6.2.1 LEAD AS AN ANTIKNOCK AGENT

In the 1920s, Thomas Midgley and his colleagues, chemists at General Motors, in order to improve the octane number, tested different compounds in gasoline. After some experiments, they discovered that tetraethyl lead (TEL), added in small quantities, allowed detonation delay and consequently generated a more complete combustion. The discovery of this compound was a success, and although it was known how harmful lead was to health and the environment, only a year later, it was industrialized by Ethyl Gasoline Corporation. This additive was added to all gasoline in the world. However, over time, it was found to be extremely harmful to health as it caused neurological problems or even fatal events, starting with the death of Ethyl GC workers; due to the foregoing, it was phased out of gasoline from the 1980s onward.[10]

It is estimated that the largest amount of lead in the environment comes from TEL added to gasoline. In a recent study by Lacerda et al.[11] the overall trend of blood lead concentration over time was evaluated. The collected data were from the 2000–2020 period. It is noticeable that there has been an exponential decrease in blood lead content since the elimination of TEL as a gasoline additive. It is clear that TEL brought great advantages in terms of economics and as an engine enhancer, but the damage to human health and the environment outweighed these benefits.

6.2.2 METAL-BASED OCTANE-BOOSTING COMPOUNDS

With the elimination of TEL, the use of other organometallic compounds was implemented to increase the antiknock quality of fuels. The most common organometallic additives added to gasoline are methylcyclopentadienyl manganese tricarbonyl (MMT), cyclopentadienyl manganese tricarbonyl (cymanthrene (CMT)), and dicyclopentadienyl iron (ferrocene (FC)).

The amount of these additives required to increase the octane number is minimal, regularly used in proportions of less than 2 wt.%. Although these compounds are believed to poorly affect the properties of fuels by the amount added, studies have shown their efficiency in retarding fuel ignition. In a study by Bruno et al.,[12] distillation curves were analyzed for gasoline blends with MMT and FC additives. It was found that as the distillate volume grows there is an increasing deviation of vaporization temperatures in blends with 40 ppm and 20 ppm FC and MMT with respect to gasoline, which is related to fuel efficiency and performance. It is also remarkable that the higher the amount of additive in the fuel, the higher the vaporization temperature.

Pitz and Westbrook[13] found that autoignition is significantly influenced by the production or consumption reactions of peroxides (HO_2 and H_2O_2) generated in the last part of combustion by oxidation of the fuel at high temperatures. In a study by Fenard et al.,[14] it was observed that the addition of FC to fuels increased the mixture ignition delay, and this fact was favored as the amount of FC in the fuel increased. It was also suggested that the inhibition of the early explosion was due to the reactions of HO_2 and H_2O_2 on the surface of iron oxide particles.

Although further research is needed to explain the inhibitory effect of organometallic compounds in gasoline, it could be assumed that this is due to the reaction of metal oxides generated during combustion, which inhibit the early explosion of the fuel caused by peroxides in the reaction mixture.

Organometallic compounds improve combustion engine efficiency and retard fuel ignition, but have some important disadvantages, for the addition of these compounds to gasoline increases CO, HC, and NO_x emissions. They also exhibit particulate emissions of metal oxides. Both iron- and manganese-based compounds generate solid deposits in the engine. These solids are deposited in the engine exhaust pipes and spark plugs, which reduce the engine's efficiency and wear it out significantly over time.

The low use of the above-mentioned additives is mainly due to the generated highly polluting emissions and their high toxicity which, although not comparable to that produced by using TEL, represents an environmental and human health issue. Prolonged exposure to manganese compounds can lead to the development of manganism, a neurodegenerative disease. Iron oxides are also harmful to health and the environment. For this reason, the use of other less-toxic compounds that increase the octane number, improve the properties of gasoline, and reduce the amount of pollutant emissions seems to be a viable solution. These compounds are called oxygenating agents.

6.3   OXYGENATING AGENTS

With the elimination of TEL in gasoline and the disuse of organometallic compounds, a need arose to search for implementing a new chemical compound that would enhance the properties of this fuel. In this sense, it was found that the addition of oxygenates to gasoline causes changes in its properties, such as octane number, heating value (LHV), Reid vapor pressure (RVP), volatility, and density, among others.

The increase in octane rating allows higher compression of gases, and growing oxygen content in the reaction mixture results in more complete combustion and significantly reduces the emission of polluting gases.[15] The calorific value decreases with increasing oxygen content in the fuel and rises with higher carbon and hydrogen content. RVP influences the cold start-up performance of the engine and is an important factor in meeting evaporative emission requirements. Volatility is the ability of a fuel to evaporate at a given temperature. If the fuel does not have adequate volatility at the correct temperature, proper combustion does not occur, and higher fuel consumption and pollutant emissions (CO and HC) take place. This property is directly related to the boiling point, where the lower the boiling point, the higher the fuel volatility. Density, on the other hand, controls the fuel atomization characteristics and energy content.[16]

The mainly employed oxygenating agents are compounds based on ethers and alcohols, due to their high RON and MON levels. Esters and carbonates have also been investigated as oxygenating agents in gasoline; however, there is still a long way to go in research and technological development to use them commercially in gasoline blends.[17–26]

6.3.1 ETHERS

Ether-based oxygenates such as methyl tert-butyl ether (MTBE), ethyl tert-butyl ether (ETBE), tert-amyl methyl ether (TAME), and diisopropyl ether (DIPE) have high-octane numbers and, therefore, are used as oxygenating agents in gasoline blending. Figure 6.3 shows the RON, MON, and OS values for the different compounds mentioned above. Since they have high octane values, they are favorable to be used in gasoline blends and improve their antiknock properties.

Although ethers have a high octane number, the improvement in gasoline properties is not always dependent on this value since some characteristics such as oxygen

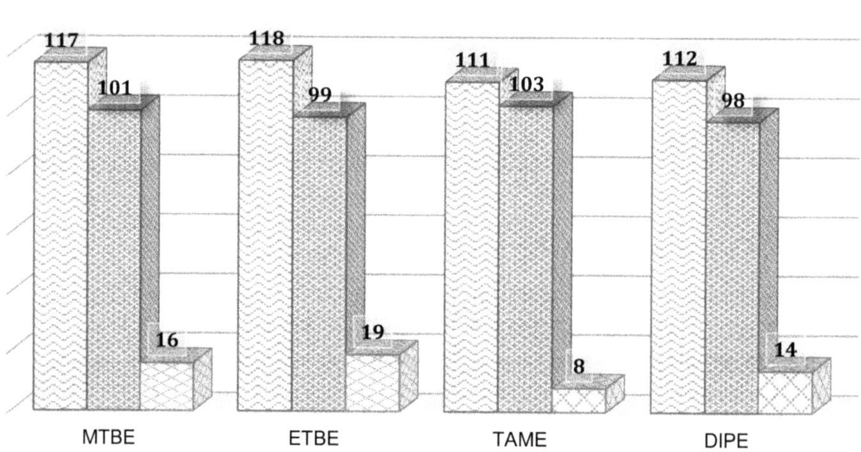

FIGURE 6.3   RON, MON, and OS values of the different ethers used as oxygenating agents (with data taken from references 9 and 17).

FIGURE 6.4 Structure of ethers used as oxygenating agents.

percentage or density are influenced by the compound chemical structure. Therefore, the qualities of the ether–gasoline mixture are always different for each added compound. Figure 6.4 shows the chemical structure of the ethers used in gasoline blends.

Ethers are used as antiknock agents due to their high calorific value, oxygen content, and low water solubility, among many others. In addition, they can be added to fuel without the need for modifications to the engine or transport pipelines. They have low vapor pressures and do not exhibit azeotropic behavior.[18] Table 6.1 contains a compilation of the physical and chemical properties of these types of gasoline oxygenating compounds.

The main compound added to gasoline as an antiknock agent after the elimination of TEL is MTBE. This compound is used because of its low cost and ease of production. It provides an increase in octane rating and a decrease in CO, HC, and NO_x emissions. Notwithstanding, studies on MTBE have reported cancer risks and environmental pollution, and for this reason, it is advisable that new alternatives be explored.[22,23]

Although TAME, ETBE, and DIPE have a very similar elemental composition (C, H, and O), they exhibit different properties and behavior patterns, which modify the specific qualities of the final gasoline blend. For example, adding ETBE to gasoline improves the thermal efficiency of the engine and causes a moderate reduction in fuel consumption. In a study by Rodriguez-Antón et al.,[24] it was found that the addition of ETBE to gasoline reduced its RVP in a manner proportional to the added ETBE volume. ETBE has also been shown to be harmless to human health;[25] thus, it is a suitable oxygenating agent to be used in gasoline, as it reduces toxic emissions from fuels and provides a high-octane number. TAME, like ETBE, has a lower RVP and a higher calorific value than MTBE, and despite its lower octane rating and difficult production that partly inhibits its use, TAME induces a lower RVP and a higher calorific value than MTBE in the final blend.[26] DIPE, on the other hand, is a suitable fuel for blending with gasoline; as it has a low RVP, it provides a high-octane number, high oxygen content, and low toxic emissions. In a study by Dhamodaran et al.,[20] it was found that adding DIPE to gasoline reduces the fuel density, RVR, and calorific value. Such a blend exhibits excellent volatility properties. In the 10% DIPE–gasoline blend, an increase of 5.67 units in RON value was obtained.

Dimethyl ether (DME) as an oxygenating agent in gasoline has been scarcely studied, but some studies have shown it to be a feasible solution for improving engine combustion and thermal efficiency, as well as decreasing harmful compounds.[27,28]

TABLE 6.1
Physical and Chemical Properties of Ether-based Oxygenating Agents

	RVP (kPa)	O (wt.%)	LHV (MJ/Kg)	Boiling Point (°C)	Auto-ignition Temperature (°C)	Flash Point (°C)	Density (g/ml) 20°C	Water Solubility (wt.%) 20°C	References
MTBE	61	18.15	34.9	52.2	443	−27	0.744	4.8	9,17,19
ETBE	20.7	15.7	36.5	72.86	310	−11	0.770	1.2	9,17,19
TAME	22	15.7	36.5	86.3	430	−19	0.764	1.15	9,17
DIPE	47	15.7	37	68	443	−28	0.726	0	9,20
DME	–	34.8	28.8	−25.1	235	−40	0.661	7	16,21

6.3.2 ALCOHOLS

The high-octane value of alcohols makes them suitable compounds for blending with gasoline. Figure 6.5 shows the RON, MON, and OS values for methanol, ethanol, different propanol isomers, butanol, and pentanol.

The oxygen content of alcohols, which is higher than that of ethers, leads to more complete combustion with fewer pollutants. Alcohol fuels are easy to store and transport due to their high self-ignition temperature and low flash point; however, alcohols have a lower calorific value. Therefore, the volume of fuel used, and combustion efficiency are compromised. Table 6.2 shows the physicochemical properties of alcohols that are of interest as gasoline enhancers.

The properties of alcohols vary depending on their alkyl chain. The percentage of oxygen present in each compound is not the same for all species; so, depending on the used alcohol, different properties are obtained in the final fuel mixtures. The structures of the alcohols are shown in Figure 6.6.

The alcohols that have been most studied in internal combustion engines have been methanol and ethanol.[30,31] The calorific value of these alcohols is lower than that of gasoline, so a considerable content of alcohol is required in gasoline to generate the same amount of energy; however, these blends can be modulated for optimum fuel efficiency.[32,33]

Methanol has a low boiling point, which leads to rapid evaporation. Its high oxygen content means better combustion of the reaction mixture, which is reflected in low pollutant emissions. In a study by Mishra et al.,[34] blends of methanol with gasoline at moderate percentages (5–15%) were analyzed. Such blends showed lower pollutant emissions than those of pure gasoline and a lower tendency to detonate, which resulted in more efficient combustion.

Ethanol has now become the most widely used oxygenating agent in gasoline blending. The addition of ethanol to gasoline, when the alcohol concentration is less

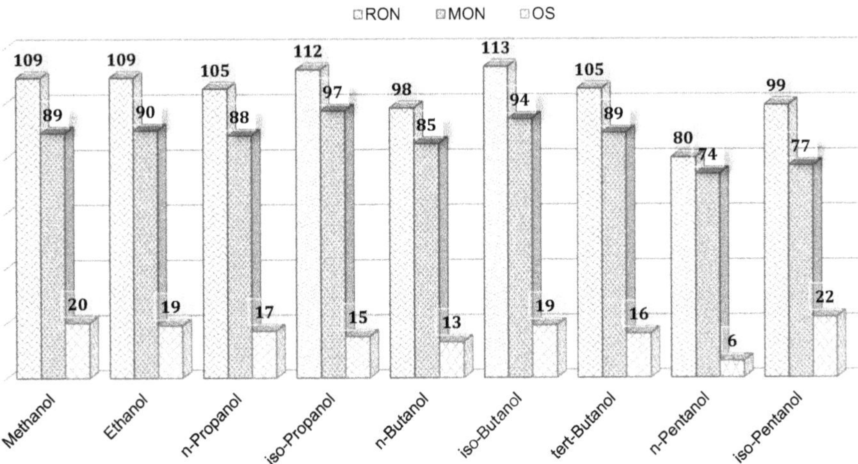

FIGURE 6.5 RON, MON, and OS values of the different alcohols used as oxygenating agents (with data obtained from Westbrook, et al.[29]).

TABLE 6.2
Physicochemical Properties of Alcohol-based Oxygenates

	RVP (kPa)	O (wt.%)	LHV (MJ/Kg)	Boiling Point (°C)	Auto-ignition Temperature (°C)	Flash Point (°C)	Density (g/cm³) 20°C	Water Solubility (wt.%) 20°C	References
Methanol	35	50	19.89	64.5	464	11	0.794	100	15,30,31
Ethanol	17	35	26.42	78.4	423	17	0.81	100	15,30,31
n-propanol	6	27	30.88	97	371	27	0.8	100	30,31
iso-propanol	15.7	27	30.54	83	399	23	0.789	100	30,31
n-butanol	8.4	21.6	33.05	117.7	343	34	0.814	7.3	15,30,31
iso-butanol	8.5	21.6	33.29	108	416	28	0.802	8.5	30,31
tert-butanol	14	21.6	25.7	82	478	16	0.787	100	30,31
n-pentanol	–	18.15	28.5	138	320	49.1	0.814	2.2	15,31,32
iso-pentanol	2.9	18.15	27.8	130	–	43	0.809	2.7	15,31,33

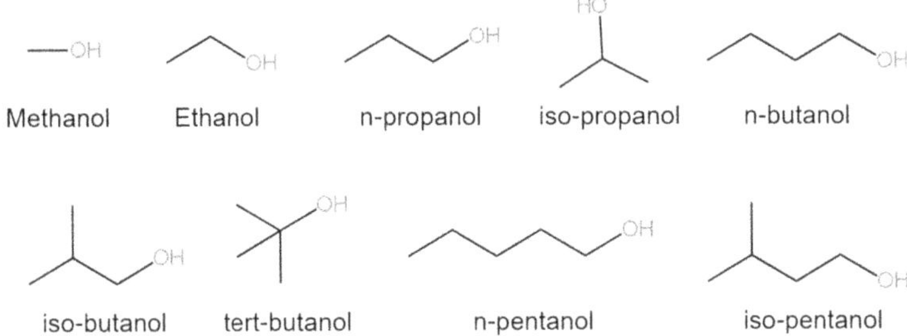

FIGURE 6.6 Structures of the alcohols used as oxygenating agents.

than 35%, increases the blend RVP. Blends with higher ethanol concentration have lower RVP than gasoline.[24] The use of ethanol reduces the tendency to detonate and improves the combustion process. In a study by Wang et al.,[35] it was shown that when ethanol is added in high proportions (>50% in the final blend), problems in cold start-up may rise. Other potential drawbacks of alcohols are their high-water miscibility, potential corrosion of metal components in the combustion system, and damage to plastic materials. In general, a satisfactory outcome is obtained with the implementation of low-molecular-weight alcohols in gasoline blends, along with their economic and easy production.

Propanol has a higher energy density than ethanol. There are two isomers of propanol, n-propanol, and iso-propanol. These isomers have not been used on a large scale because they are more expensive to obtain than ethanol. Masum et al.,[36] found that 20% n-propanol/gasoline blends decrease CO and HC emissions compared to pure gasoline and increase its thermal efficiency. In another research work, a decrease in CO and HC emissions was also observed in iso-propanol/gasoline blends at moderate proportions (5–10%).[37]

Alcohols with higher molecular weight than ethanol and propanol, that is, n-butanol, offer higher energy density, lower azeotropic behavior, lower corrosivity of engine metal compounds, and water miscibility than lower-molecular-weight alcohols. They also show higher LHV and lower NO_x emissions, among many other advantages that make them promising additions to gasoline blends.[37] In a study conducted by Elfasakhany,[38] it was observed that in butanol with gasoline blends, polluting emissions of CO and HC decreased drastically, while the power and volumetric efficiency of the fuel decreased to a lesser extent.

N-butanol, iso-butanol, and tert-butanol can be used as oxygenating agents in gasoline due to their specific chemical properties. In a study by Gu et al.,[39] n-butanol blends were used with gasoline at different percentages (10, 30, 40, and 100%); in this study, it was shown that as the n-butanol content in the fuel increased, the fuel consumption also grew, which was mainly related to its high calorific value. On the other hand, if pure n-butanol was used (100%), tailpipe emissions increased. As it has a lower octane number than ethanol and methanol, butanol requires a lower compression ratio, which leads to lower efficiency and, consequently, to higher pollutant emissions. Iso-butanol improves fuel conversion efficiency by 6% when using a 50% iso-butanol/gasoline blend and the fuel conversion efficiency decreases when using 100% iso-butanol.[40] Tert-butanol, unlike the other butanol isomers, has a higher RVP than gasoline. It has also been shown that tert-butanol increases the RON, MON, and LHV in blends at varied proportions with gasoline.[41]

Pentanol isomers, n-pentanol and iso-pentanol, have lower octane numbers than gasoline and therefore are not as widely used in internal combustion engines as lower-molecular-weight alcohols. Despite this, studies have shown their efficiency in gasoline mixtures. In a recent study,[42] an investigation was conducted on the engine performance and emissions of n-pentanol blends with gasoline at moderate percentages (5–20%). All the analyzed samples exhibited improved engine performance compared to pure gasoline; the results also showed a reduction in CO, HC, and NO_x emissions. The decrease in NO_x emissions was attributed to the high LHV of n-pentanol.

Undoubtedly, the use of alcohols in gasoline blends is a feasible alternative to improve combustion processes, increase the gasoline octane rating, improve engine performance, and decrease the amount of pollutant emissions.

6.3.3 ESTERS

Esters are compounds that contain two oxygen atoms in their structure, so they are also used as oxygenating agents. The advantage of their use is the increase in the octane number of gasoline without changes in the RVP values.[43] They are nontoxic and do not produce harmful gases such as aldehydes, ketones, and CO. Additionally, they have low volatility, ease of handling, and relatively low production cost compared to other gasoline enhancers. Different types of esters like methyl acetate, ethyl acetate, ethyl propionate, and ethyl butyrate have been studied. The structures of the esters are shown in Figure 6.7.

In one study,[43] analyses of gasoline blends with methyl acetate and ethyl acetate were performed. In both cases, an increase in RON values was observed as the percentage of ester in the fuel increased. In another report,[44] a sampling of ethyl acetate blends with petrol at 4, 8, and 12% by volume was carried out. There, it was observed that as the amount of ethyl acetate in the fuel was higher, the fuel consumption also grew. It was also found that all the samples decreased the CO, HC, and NO_x emissions at the expense of an increase in CO_2 emissions. Badway et al.[45] conducted a study of different esters, such as methyl acetate, ethyl acetate, ethyl propionate, and ethyl butyrate, showing that ethyl propionate and ethyl butyrate exhibited better combustion characteristics than those of ethanol and gasoline.

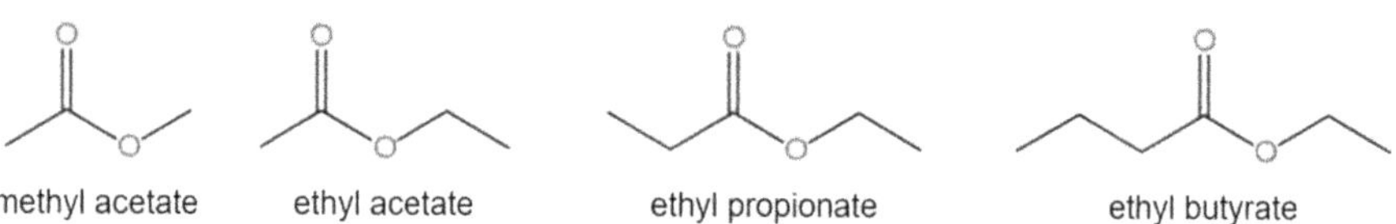

FIGURE 6.7   Structures of the esters used as oxygenating agents.

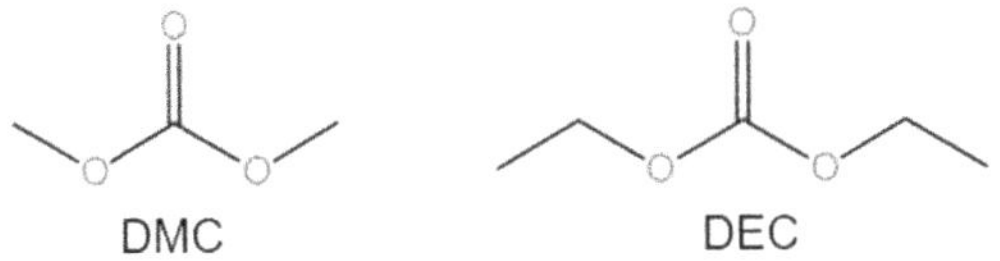

DMC   DEC

FIGURE 6.8 Structures of carbonates used as oxygenating agents.

6.3.4 CARBONATES AND ACETATES

Carbonates have also received attention as oxygenating agents. Dimethyl carbonate (DMC) and diethyl carbonate (DEC) are both biodegradable and nontoxic, have high-octane numbers and high oxygen content, and are not miscible in water like some alcohols. Figure 6.8 shows the chemical structures of the carbonates used in gasoline blends.

In a study conducted by Pacheco and Marshal,[46] it was reported that DMC had the potential to be added to gasoline as an oxygenating agent due to its low RVP, high octane number, high density, and decreased CO and NO_x emissions. In another study,[47] an 8% DMC–gasoline blend was evaluated, showing a 30% reduction in unburned HCs and a 60% decrease in the total number of emissions. The thermal efficiency of the engine remained the same with the 8% blend as with the pure fuel.

Another way to obtain oxygenating agents for gasoline is to transform the waste obtained in the synthesis of biodiesel.[48] Crude glycerol is the raw material to be transformed into glycerol acetates (GAs). Through single-cylinder equipment tests, RON and MON were measured and octane number results were obtained as an average. The study included the use of blended ethanol, MTBE, and GAs at concentrations of 5 and 10 wt.% of each oxygenate. The conclusions of this study were that these novel oxygenating agents had a similar octane number behavior pattern to that one identified when using ethanol or MTBE in their respective proportions. The use of GAs for this purpose has been then demonstrated.

6.4 REDUCTION OF GASEOUS EMISSIONS WITH THE USE OF OXYGENATING AGENTS IN PETROL

As mentioned above, since the 1970s, the world's major refiners began to phase out TEL as an antiknock agent because of the problems that lead causes to human health and all living matter. Lead toxicity causes serious health problems, such as kidney failure, cardiac arrhythmias, and brain and lung damage.

The substitution of lead by antiknock agents is aimed at maintaining the properties of gasoline, increasing the octane number, and decreasing the aromatic content of fuels. The desirable composition of oxygenating components in gasoline is around 3–5 wt.%, which has been shown to reduce the pollutants emitted by the engine, thus improving the air quality.[49] It is important that the amount of such oxygenating additives be controlled as they can reduce emissions of pollutants such as CO and HCs, but they may emit toxic combustion components like aldehydes and peroxyacetyl nitrates.[50]

TABLE 6.3

Gasoline-oxygenated Composition and its Relationship with Exhaust Emissions

Fuel	Emissions	References
Gasoline-butanol(2.5)-water(19)	Reduces NO_x emissions under all operating conditions	[51]
Gasoline-ethanol(15)-water(10–30)	Reduces 7.6% of HC and 28% of NO_x	[52]
Gasoline-ethanol plus natural gas compressed by direct injection	Reduces HC by 11–23 %.	[53]
Gasoline-ethanol(50)	HC, CO, and NO_x emissions decrease by 27%, 33%, and 29%, respectively.	[54]
Gasoline-ammonia/hydrogen (5)	NO_x emissions increase by 42%.	[55]

An interesting alternative is to use oxygenating agents obtained from biomass, such as alcohols that can provide 10–45% oxygen.[51] A summary of some oxygenating agent options and their relationship with atmospheric emissions is presented in Table 6.3. It is observed that emissions of HCs, CO, and NO_x decrease in all cases with the addition of oxygenating agents.[52-54] As for the addition of ammonia/hydrogen,[55] although the energy efficiency is improved, an increase in NO_x is observed; however, this fact is beyond the scope of this chapter.

Combustion is a chemical reaction that under all engine operating conditions, even in excess oxygen, proceeds to equilibrium and is always an incomplete reaction. Chemical equilibrium is defined as a dynamic state, where the rate of appearance of the products is equal to the rate of disappearance of the reactants. This means that there will always remain a quantity of unburned noxious gases in the engine cylinders. The minimum quantities of these pollutant gas emissions depend on the air–fuel ratio and the operating temperature of the combustion process. On the one hand, CO and HC are the products of incomplete combustion, that is oxygen-deficient combustion, while the presence of high amounts of NO_x occurs at high operating temperatures and high amounts of oxygen in the feed mixture. Emissions of gaseous pollutants from internal combustion engines, whether using gasoline or diesel, are contributing to global atmospheric changes and are dangerous for all living beings and environmental balance.[51]

The addition of oxygenating agents to gasoline improves the combustion process and, therefore, fewer pollutants are emitted into the atmosphere. However, they should be used sparingly in order to avoid exceeding the recommended values of 3–5 wt.% as it may affect other fuel properties such as viscosity, flash point, volatility, and calorific value, among others. It could also modify important properties such as the combustion rate and, thus, the rate of heat release, which would be counterproductive to their use as it would lead to incomplete combustion. The combination of oxygenating agents and air in the right proportions improves fuel quality by increasing the octane and cetane number of gasoline and diesel fuels, respectively. Additionally, increasing the temperature of engine cylinders decreases the proportion of particulate pollutants (>10 μm).[56] Despite the advantages described with the use of oxygenating agents, it is important to note that the combustion of alcohols and ethers produces aldehyde emissions.[56] These compounds are classified as cancer precursors and, from an environmental point of view, lead to the generation of ozone in the troposphere, and it is therefore necessary to improve engine and exhaust technologies in order to have the most complete combustion possible through their optimization. Despite the good properties of MTBE in the combustion process of gasoline, it is recommended to use it sparingly. This organic compound has been found to be carcinogenic to humans and animals and, in case of spills or sneaky emissions, it can contaminate groundwater. It is recommended for use in amounts not exceeding 3 wt.%.[56]

Another point to be considered in the use of alcohols as oxygenating agents is that they have a high vapor pressure that could contribute to fugitive emissions if proper controls are not in place. In addition, another point against them is that they do not have adequate values (3–5 wt.%) for cold start-up and in that sense, they have disadvantages with respect to other oxygenating agents with higher latent heat of vaporization. Additionally, it is necessary to consider the behavior of metals and elastomeric components of the ducts that carry the fuel to the engine, which must be redesigned in their composition in order to have a correct useful life with the use of alcohols.[57] The problems of cold start-up or even methanol segregation due to the presence of water can be solved by using higher-molecular-weight alcohols, such as tert-butanol or iso-butanol; unfortunately, these compounds do not have the advantages of local production, such as that of ethanol, which can even be obtained from agro-industrial waste.

6.5 LIMITATIONS AND DISADVANTAGES OF USING ALCOHOL-BASED OXYGENATING AGENTS

Despite the many reported benefits of fuels produced by blending alcohols with gasoline, including increased octane rating and reduced emissions of pollutants such as NO_x, CO, and HC, it is not possible to increase the massive use of this oxygenating agent in all regions of the world, since there is no mass production of it. In order to have an orderly transition with oxygenating agents, it is necessary to have a wide diversity of molecules to improve combustion efficiency and to encourage alternative markets.

An important point to note is that not all organic molecules that have oxygen in their structure can be considered as oxygenating agents for fuels. The ignition quality and the source of renewable feedstock should not be the only parameters to determine the viability of oxygenating agents. For example, Gouli et al.[58] reported on the behavior of furan and p-cresol derivatives as oxygenating agents with suitable antiknock properties and a decrease in volatile HC and CO emissions, which are a function of the air–fuel ratio. The evaluated compounds were 2-methylfuran, furfuryl alcohol, furfurylamine, and p-cresol. However, the processes to obtain furan make it technically unaffordable and unavailable to the market.

There are also studies of little industrial feasibility that use blends of edible oils with diesel to obtain improvements in combustion processes, in terms of both mechanical behavior and improved vehicle emissions. Niemi et al.[59] proposed the use of cold-pressed oil from the seeds of brown mustard (*Brassica juncea*). While it is true that there is an increasing trend toward the use of biofuels or their derivatives in the fossil fuel market, it is also important to avoid the use of edible raw materials for fuel production. This should even be part of the philosophical analyses of how natural resources for human food should be used in relation to the need for transport. From a purely technical point of view, it is observed that the use of mustard oil produces less smoke and particulate matter in the exhaust emissions of a diesel engine for off-road vehicles.[59] The mixing procedure of oils together with diesel was not problematic and was well integrated over the whole range of studied concentrations.

There are three processes for obtaining alcohols and ethers from biomass.[60] The first one is bacterial fermentation (yeast). The second one is gasification from ligno-cellulosic residues in which a mixture of alcohols is obtained. The third one is acid hydrolysis, which produces 4- or 5-carbon alcohols. An important advantage of alcohols is that their production can originate from biomass as well as from industrial chemical processes. However, their production from agricultural and/or forestry biomass represents a very heavy environmental burden due to the high requirements of fertile land and water consumption necessary for their production. In addition, the water content in biofuels decreases the calorific value of the fuel based on low-molecular-weight alcohols, making it necessary to use larger quantities of fuel to provide the energy requirements of engines.[58] Therefore, some authors do not consider it to be a comprehensive and feasible solution to the use of fossil fuels. Another disadvantage of this type of fuel is the low boiling point and high vapor pressures of both alcohols and ethers, so it is necessary to develop additional devices to adequately contain their volatility. Additionally, they cause corrosion in some engine parts, both metallic and polymeric.

Due to the hydrophilic capacity of alcohols, a fuel mixture containing them is really a three-component system: water–alcohol–HCs. The lower the molecular weight of the alcohol, the greater the problems associated with phase separation. The most suitable of alcohols to be integrated into fuels seems to be butanol.

In addition, some alcohols have low viscosity that causes leaks during injection and pumping. So, current engine technology would require adaptations to use fuel with high concentrations of alcohols.[61]

6.6 CONCLUSIONS AND OUTLOOK

The addition of compounds that improve the properties of gasoline has become a fundamental part of the correct performance and operation of the internal combustion engine. The development of these compounds has gone through a long road of experimentation and research. Initially, a compound was obtained that greatly improved the antiknock properties of gasoline, but it was also the cause of catastrophic results, causing the death of thousands of people and environmental pollution, which, although it has been reduced over the years, persists today. Similar cases occurred with other organometallic compounds since the number of polluting emissions is worrying and the efficiency they provide to the internal combustion engine is scarce.

Undoubtedly, oxygenating agents have great potential as octane boosters, since, as we have seen in this chapter, they reduce the amount of harmful exhaust gases, improve engine performance and, due to the amount of oxygen contained in them, provide efficient combustion. However, not all of them are recommended, as is the case of MTBE, which generates toxic emissions into the environment, or low-molecular-weight alcohols, which, although they are accessible and easy to produce, generate corrosion in the engine metallic and polymeric systems.

At this point, it becomes essential to address three important aspects: the choice of the octane enhancer with the best qualities in gasoline, the accessibility of these compounds, and the possible modifications of the combustion engine.

It is clear that the most balanced compound in terms of performance and emissions must be chosen. All this, guided by the properties of the different compounds, can affect or benefit gasoline to a greater or lesser extent. The compound must provide high miscibility in gasoline, increase compression capacity, and modulate RVP and volatility, among others. The balance in these properties is the key point to obtain an adequate octane-boosting compound, usable in gasoline blends.

The accessibility of octane-boosting compounds is a very important point to be considered. Although some compounds such as high-molecular-weight alcohols exhibit excellent properties in gasoline blends, their production is highly costly and difficult to scale up. Therefore, the implementation of this type of compound is almost economically inaccessible. The lack of accessibility to the compounds and the need to make modifications to the engine go hand in hand. The development of technologies is an indispensable requirement, since either modifications of the combustion engine are required to avoid corrosion problems and gasket wear with the use of oxygenating compounds or adequate technologies must be provided to facilitate access to compounds that greatly benefit combustion, reduce emissions, increase engine performance, and improve the antiknock gasoline capacity.

REFERENCES

1 Monaghan, M. L.; "Chapter one-Introduction," *Internal Combustion Engines*, Constantino Arcoumanis, Academic press, 1988, pp 1–30. doi: 10.1016/B978-0-12-059790-1.50007-9

2 Heywood, J. B.; *Internal Combustion Engine Fundamentals*, 2nd ed, New York, McGrawHill, 1988, pp 8–20.

3 Corrubia, J. A.; Capece, J. M.; Cernansky, N. P.; Miller, D. L.; Durrett, R. P.; Najt, P.M. RON and MON chemical kinetic modeling derived correlations with ignition delay time for gasoline and octane boosting additives, *Combust Flame*, 2020, 219, 359–372. doi: 10.1016/J.COMBUSTFLAME.2020.05.002

4 Westbrook, C. K.; Sjöberg, M.; Cernansky, N. P. A new chemical kinetic method of determining RON and MON values for single component and multicomponent mixtures of engine fuels, *Combust Flame*, 2018, 195, 50–62. doi: 10.1016/J.COMBUSTFLAME.2018.03.038

5 Polikarpov, E.; Bays, J. T.; Lilga, M. A.; Guo, M. F.; Gaspar, D. J. The effect of chemical functional groups on the octane sensitivity of fuel blends for spark-ignited and multimode engines, *Fuel*, 2023, 352, 129107. doi: 10.1016/J.FUEL.2023.129107

6   Badra, J. A.; Bokhumseen, N.; Mulla, N.; Sarathy, S. M.; Farooq, A.; Kalghatgi, G.; Gaillard, P. A methodology to relate octane numbers of binary and ternary n-heptane, iso-octane and toluene mixtures with simulated ignition delay times, *Fuel*, 2015, 160, 458–469. doi: 10.1016/J.FUEL.2015.08.007

7   Athafah, N.; Salih, A. Effect of octane number on performance and exhaust emissions of an SI engine, *Eng Technol J*, 2020, 38, 574–585. doi: 10.30684/ETJ.V38I4A.263

8   Sayin, C.; Kilicaslan, I.; Canakci, M.; Ozsezen, N. An experimental study of the effect of octane number higher than engine requirement on the engine performance and emissions, *Appl Therm Eng*, 2005, 25, 1315–1324. doi: 10.1016/J.APPLTHERMALENG.2004.07.009

9   Abdellatief, T. M. M.; Ershov, M. A.; Kapustin, V. M.; Ali Abdelkareem, M.; Kamil, M.; Olabi, A. G. Recent trends for introducing promising fuel components to enhance the anti-knock quality of gasoline: A systematic review, *Fuel*, 2021, 291, 120112. doi: 10.1016/J.FUEL.2020.120112

10 Splitter, D.; Pawlowski, A.; Wagner, R. A historical analysis of the co-evolution of gasoline octane number and spark-ignition engines, *Front Mech Eng*, 2016, 1, 160787. doi: 10.3389/FMECH.2015.00016/BIBTEX

11 Lacerda, D.; Pestana, I. A.; Santos Vergilio, C.; Rezende, C. E. Global decrease in blood lead concentrations due to the removal of leaded gasoline, *Chemosphere*, 2023, 324, 138207. doi: 10.1016/J.CHEMOSPHERE.2023.138207

12 Bruno T. J.; Baibourine, E. Analysis of Organometallic Gasoline Additives with the Composition-Explicit Distillation Curve Method, *Energy Fuels*, 2010, 24, 5508–5513. doi: 10.1021/EF1006403

13 Pitz, W. J.; Westbrook, C. K. Chemical kinetics of the high pressure oxidation of n-butane and its relation to engine knock, *Combust Flame*, 1986, 63, 113-133, 0010–2180. doi: 10.1016/0010-2180(86)90115-X

14 Fenard, Y.; Song, H.; Dauphin, R.; Vanhove, G. An engine-relevant kinetic investigation into the anti-knock effect of organometallics through the example of ferrocene, *Proc Combust Inst*, 2019, 37, 547–554,1540–7489. doi: 10.1016/j.proci.2018.06.135

15 Das, A. K.; Sahu, S. K.; Panda, A. K. Current status and prospects of alternate liquid transportation fuels in compression ignition engines: A critical review, *Renewable Sustainable Energy Rev*, 2022, 161, 112358. doi: 10.1016/J.RSER.2022.112358

16 Shrivastav, G.; Ahmad, E.; Khan, T. S.; Ali Haider, M. Customizing reformulated gasoline using biofuel-additives to replace aromatics, *J Mol Liq*, 2024, 398, 124251. doi: 10.1016/J.MOLLIQ.2024.124251

17 Neagu Petre, M.; Rosca, P.; Dragomir, R. E. The Effect of Bio-ethers on the Volatility Properties of Oxygenated Gasoline, *Internet REV. CHIM.* [Online] 2011, 62, http://www.revistadechimie.ro567 (accessed: Apr. 17, 2024).

18 Wallace, G.; Blondy, J.; Mirabella, W.; Schulte-Körne, E.; Viljanen, J. Ethyl tertiary butyl ether - A review of the technical literature, *SAE Int J Fuels Lubr*, 2009, 2, 940–952. doi: 10.4271/2009-01-1951

19 Surisetty, V. R.; Dalai, A. K.; Kozinski, J. Alcohols as alternative fuels: An overview, *Appl Catal A Gen*, 2011, 404, 1–11. doi: 10.1016/J.APCATA.2011.07.021

20 Dhamodaran G.; Esakkimuthu, G. S. Experimental measurement of physico-chemical properties of oxygenate (DIPE) blended gasoline, *Measurement*, 2019, 134, 280–285. doi: 10.1016/J.MEASUREMENT.2018.10.077

21 Verbeek, R.; Van Der Weide, J. Global assessment of dimethyl-ether: Comparison with other fuels, *SAE Technical Papers*, 1997. doi: 10.4271/971607

22 Hartley, W. R.; Englande, A. J.; Harrington, D. J. Health risk assessment of groundwater contaminated with methyl tertiary butyl ether (MTBE), *Water Sci Technol*, 1999, 39, 305–310. doi: 10.1016/S0273-1223(99)00290-5

23 Squillace, P. J.; Pankow, J. F.; Korte, N. E.; Zogorski, J. S. Review of the environmental behavior and fate of methyl tert-butyl ether, *Environ Toxicol Chem*, 1997, 16, 1836–1844. doi: 10.1002/ETC.5620160911

24 Rodríguez-Antón, L. M.; Hernández-Campos, M.; Sanz-Pérez, F. Experimental determination of some physical properties of gasoline, ethanol and ETBE blends, *Fuel*, 2013, 112, 178–184. doi: 10.1016/J.FUEL.2013.04.087

25 Medinsky, M. A.; Wolf, D. C.; Cattley, R. C.; Wong, B.; Janszen, D. B.; Farris, G. M. Effects of a thirteen-week inhalation exposure to ethyl tertiary butyl ether on Fischer-344 rats and CD-1 mice, *Toxicol Sci*, 1999, 51, 108–118. doi: 10.1093/toxsci/51.1.108

26 Ershov, M. A.; Potanin, D. A.; Tarazanov, S. V.; Abdellatief, T. M. M.; Kapustin, V. M. Blending Characteristics of Isooctene, MTBE, and TAME as Gasoline Components, *Energy Fuels*, 2020, 34, 2816–2823. doi: 10.1021/ACS.ENERGYFUELS.9B03914/SUPPL_FILE/EF9B03914_SI_001.PDF

27 Liang, C.; Ji, C.; Liu, X. Combustion and emissions performance of a DME-enriched spark-ignited methanol engine at idle condition, *Appl Energy*, 2011, 88, 3704–3711. doi: 10.1016/J.APENERGY.2011.04.056

28 Liang, C.; Ji, C.; Gao, B.; Liu, X.; Zhu, Y. Investigation on the performance of a spark-ignited ethanol engine with DME enrichment, *Energy Convers Manag*, 2012, 58, 19–25. doi: 10.1016/J.ENCONMAN.2012.01.004

29 Westbrook, C. K.; Mehl, M.; Pitz, W. J.; Sjöberg, M. Chemical kinetics of octane sensitivity in a spark-ignition engine, *Combust Flame*, 2017, 175, 2–15. doi: 10.1016/J.COMBUSTFLAME.2016.05.022

30 Abdellatief, T. M. M.; Ershov, M. A.; Kapustin, V. M.; Ali Abdelkareem, M.; Kamil, M.; Olabi, A. G. Recent trends for introducing promising fuel components to enhance the anti-knock quality of gasoline: A systematic review, *Fuel*, 2021, 291, 120112. doi: 10.1016/J.FUEL.2020.120112

31 Sarathy, S. M.; Oßwald, P.; Hansen, N.; Kohse-Höinghaus, K. Alcohol combustion chemistry, *Prog Energy Combust Sci*, 2014, 44, 40–102. doi: 10.1016/J.PECS.2014.04.003

32 Das, A. K.; Sahu, S. K.; Panda, A. K. Current status and prospects of alternate liquid transportation fuels in compression ignition engines: A critical review, *Renewable Sustainable Energy Rev*, 2022, 161, 112358. doi: 10.1016/J.RSER.2022.112358

33 Masum, B. M.; Masjuki, H. H.; Kalam, M. A.; Palash, S. M.; Habibullah, M. Effect of alcohol–gasoline blends optimization on fuel properties, performance and emissions of a SI engine, *J Clean Prod*, 2015, 86, 230–237. doi: 10.1016/J.JCLEPRO.2014.08.032

34 Mishra, P. C.; Gupta, A.; Kumar, A.; Bose, A. Methanol and petrol blended alternate fuel for future sustainable engine: A performance and emission analysis, *Measurement*, 2020, 155, 107519. doi: 10.1016/J.MEASUREMENT.2020.107519

35 Wang, C.; Zeraati-Rezaei, S.; Xiang, L.; Xu, H. Ethanol blends in spark ignition engines: RON, octane-added value, cooling effect, compression ratio, and potential engine efficiency gain, *Appl Energy*, 2017, 191, 603–619. doi: 10.1016/J.APENERGY.2017.01.081

36 Masum, B. M.; Kalam, M. A.; Masjuki, H. H.; Palash, S. M.; Fattah, I. M. R. Performance and emission analysis of a multi cylinder gasoline engine operating at different alcohol-gasoline blends, *RSC Adv*, 2014, 4, 27898–27904. doi: 10.1039/C4RA04580G

37 Altun, Ş.; Öner, C.; First, M. Exhaust emissions from a spark-ignition engine operating on iso-propanol and unleaded gasoline blends, *Internet Technol.* [Online] 2010, 13, 183–188, http://jestech.karabuk.edu.tr/arsiv/2010-03/PDF/7%20-%20%C5%9E.Altun.pdf (accessed: Apr. 23, 2024).

38 Elfasakhany, A. Experimental study on emissions and performance of an internal combustion engine fueled with gasoline and gasoline/n-butanol blends, *Energy Convers Manag*, 2014, 88, 277–283. doi: 10.1016/J.ENCONMAN.2014.08.031

39 Gu, X.; Huang, Z.; Cai, J.; Gong, J.; Wu, X.; Lee, C. F. Emission characteristics of a spark-ignition engine fuelled with gasoline-n-butanol blends in combination with EGR, *Fuel*, 2012, 93, 611–617. doi: 10.1016/J.FUEL.2011.11.040

40 Irimescu, A. Performance and fuel conversion efficiency of a spark ignition engine fueled with iso-butanol, *Appl Energy*, 2012, 96, 477–483. doi: 10.1016/J.APENERGY.2012.03.012

41 Niass, T.; Amer, A. A.; Xu, W.; Vogel, S. R.; Krebber-Hortmann, K.; Brassat, A. Butanol blending -a promising approach to enhance the thermodynamic potential of gasoline -part 1, *SAE Int J Fuels Lubr*, 2012, 5, 265–273. doi: 10.4271/2011-01-1990

42 Yaman H.; Yesilyurt, M. K. The influence of n-pentanol blending with gasoline on performance, combustion, and emission behaviors of an SI engine, *Eng Sci Technol, Int J*, 2021, 24, 1329–1346. doi: 10.1016/J.JESTCH.2021.03.009

43 Dabbagh, H. A.; Ghobadi, F.; Ehsani, M. R.; Moradmand, M. The influence of ester additives on the properties of gasoline, *Fuel*, 2013, 104, 216–223. doi: 10.1016/J.FUEL.2012.09.056

44 Yeşilyurt, M. K.; Erol, D.; Yaman, H.; Doğan, B. Effects of using ethyl acetate as a surprising additive in SI engine pertaining to an environmental perspective, *Int J Environ Sci Technol*, 2022, 19, 9427–9456. doi: 10.1007/S13762-021-03706-3/FIGURES/9

45 Badawy, T.; Williamson, J.; Xu, H. Laminar burning characteristics of ethyl propionate, ethyl butyrate, ethyl acetate, gasoline and ethanol fuels, *Fuel*, 2016, 183, 627–640. doi: 10.1016/J.FUEL.2016.06.087

46 Pacheco M. A.; Marshall, C. L.; Review of dimethyl carbonate (DMC) manufacture and its characteristics as a fuel additive, *Energy Fuels*, 1997, 11, 2–29. doi: 10.1021/EF9600974/ASSET/IMAGES/LARGE/EF9600974FA01N.JPEG

47 Chan J. H.; Tsolakis, A.; Herreros, J. M.; Kallis K. X.; Hergueta C.; Sittichompoo, S.; Bogarra, M. Combustion, gaseous emissions and PM characteristics of Di-Methyl Carbonate (DMC)-gasoline blend on gasoline Direct Injection (GDI) engine, *Fuel*, 2020, 263, 116742. doi: 10.1016/J.FUEL.2019.116742

48 Herrada-Vidales, J.A.; Garcia-Gonzalez, J.M.; Martínez-Palou, R. Guzman-Pantoja, J. Integral process for obtaining acetins from crude glycerol and their effect on the octane index. *Chem. Eng. Commun.*, 2019, https://doi.org/10.1080/00986445.2019.1578758

49 Ershov, M.A.; Savelenko, V.D.; Shvedova, N.S.; Kapustin, V.M.; Abdellatief, T.M.M.; Karpov, N.V.; Dutlov, E.V.; Borisanov, D.V. An evolving research agenda of merit function calculations for new gasoline compositions. *Fuel*, 2022, 322, 124209. https://doi.org/10.1016/j.fuel.2022.124209

50 Magnusson, R.; Nilsson, C. The influence of oxygenated fuels on emissions of aldehydes and ketones from a two-stroke spark ignition engine. *Fuel*, 2011, 90(3), 1145–1154. https://doi.org/10.1016/j.fuel.2010.10.026

51 Dhamodaran, G.; Esakkimuthu, G. S.; Palani, T.; Sundaraganesan, A.; Reducing gasoline engine emissions using novel bio-based oxygenates: a review, *Emergent Materials*, 2023, 6:1393–1413, https://doi.org/10.1007/s42247-023-00470-7

52 Cesur, I. Effects of water injection strategies on performance and exhaust emissions in an ethanol fueled gasoline engine. *Therm. Sci. Eng. Prog.*, 2022, 101397. https://doi.org/10.1016/j.tsep.2022.101397

53 Zhao, Z.; Huang, Y.; Yu, X.; Guo, Z.; Yu, L.; Meng, S.; Li, D. Experimental study on combustion and emission of an SI engine with natural gas/ethanol combined injection. *Fuel*, 2022, 318, 123476. https://doi.org/10.1016/j.fuel.2022.123476

54 Kaya, G. Experimental comparative study on combustion, performance and emissions characteristics of ethanol-gasoline blends in a two stroke uniflow gasoline engine. *Fuel*, 2022, 317, 120917. https://doi.org/10.1016/j.fuel.2021.120917

55 Dinesh, M.H.; Kumar, G.N. Effects of Compression and mixing ratio on NH_3/H_2 fueled si engine performance, combustion stability, and emission. *Energy Convers. Manag.: X*; 2022, 100269. https://doi.org/10.1016/j.ecmx.2022.100269

56 Tabatabaei, M.; Aghbashlo, M.; Najafi, B.; Hosseinzadeh-Bandbafha, H.; Faizollahzadeh Ardabili, S.; Akbarian, E.; Khalife, E.; Mohammadi, P.; Rastegari, H.; Ghaziaskar, H.S. Environmental impact assessment of the mechanical shaft work produced in a diesel engine running on diesel/biodiesel blends containing glycerol-derived triacetin. *J. Clean. Prod.*, 2019, 223, 466–486. https://doi.org/10.1016/j.jclepro.2019.03.106

57 Amaral, L. V.; Santos, N. D. S. A.; Roso, V. R.; Sebastiao, R. C. O.; Pujatti, J. P. Effects of gasoline composition on engine performance, exhaust gases and operational costs, *Renewable Sustainable Energy Rev*, 2021, 135, 110196. https://doi.org/10.1016/j.rser.2020.110196

58 Gouli, S.; Lois, E.; Stournas, S.; Effects of Some Oxygenated Substitutes on Gasoline Properties, Spark Ignition Engine Performance, and Emissions, *Energy Fuels*, 1998, 12, 918.

59 Niemi, S. A.; Tyrväinen, J.M.; Laurén, M.J.; Laiho, V.O.K.; Ekolaiho, O. Exhaust emissions of an off-road diesel engine driven with a blend of diesel, *Proceedings of ICES03 Spring Technical Conference of the ASME Internal Combustion Engine Division.* Salzburg, Austria, 2003, 11–14. Downloaded From: https://proceedings.asmedigital collection.asme.org/

60 Awad, O. I.; Mamaa, R.; Ibrahim, T.K.; Hammid, A.T.; Yusri, I.M.; Hamidi, M.A.; Humada, A.M.; Yusop, A.F. Overview of the oxygenated fuels in spark ignition engine: Environmental and performance, *Renew. Sust. Ener. Rev.*, 2018, 91, 394–408. https://doi.org/10.1016/j.rser.2018.03.107

61 Amaral, L.V.; Santos, N.D.; Roso, V.R.; Sebastiao, R.C.O.; Pujatti, F.J.P. Effects of gasoline composition on engine performance, exhaust gases and operational costs, *Renewable Sustainable Energy Rev.*, 2021, 135, 110196, https://doi.org/10.1016/j.rser.2020.110196

7 Gasoline from Biomass and Non-petroleum Sources

Vianey Edith López Pérez and
Mercedes Bazaldua Domínguez
Mexican Petroleum Institute, Mexico City, México

7.1 INTRODUCTION: BIOMASS AS A FEEDSTOCK

Biomass is any type of organic matter (forest residues, agricultural waste, animal manure, crop residues, algae, microorganisms, food waste, etc.) that originates from a biological process and is a renewable energy source.[1] Its main constituents are hemicellulose, cellulose, and lignin, among other extractable components.[2]

The value of biomass depends on its chemical and physical properties, which are closely related to the chemical composition of the plant or waste from which it comes and the environmental setting where the crop was grown. Biomass from thermal, chemical, and biochemical processes can be transformed into biofuels. Figure 7.1 shows the compositions of some biomass types for conversion into biofuels.[3]

Specifically, the process of generating gasoline from biomass requires multiple steps, including the preparation of the raw material by gasification, pyrolysis, or liquefaction. These processes are carried out under controlled conditions. Generally, they are conducted in an anaerobic environment because, in the production of gasoline, the presence of oxygen causes the combustion of the biomass organic compounds.[4]

The biomass-based industry is analogous to that of oil refining and distillation and is currently generating multiple fuels and products equivalent to those obtained from oil. This new concept, called biorefinery, integrates biomass conversion processes and equipment to produce biofuels (bioethanol, biomethanol, biodiesel, biohydrogen, biogas, biogasolines, etc.) to name a few.

7.1.1 SUITABLE BIOMASS TYPES FOR GASOLINE PRODUCTION

Depending on the feedstock from which they are derived, biofuels are classified into four generations (Figure 7.2).[5–7] First-generation biofuels are produced from lignocellulosic biomass from crops that can be used for human consumption; second-generation biofuels are nonfood items and the main raw material source is agricultural and forestry residues. Third-generation biofuels, unlike first- and second-generation

DOI: 10.1201/9781003517283-7

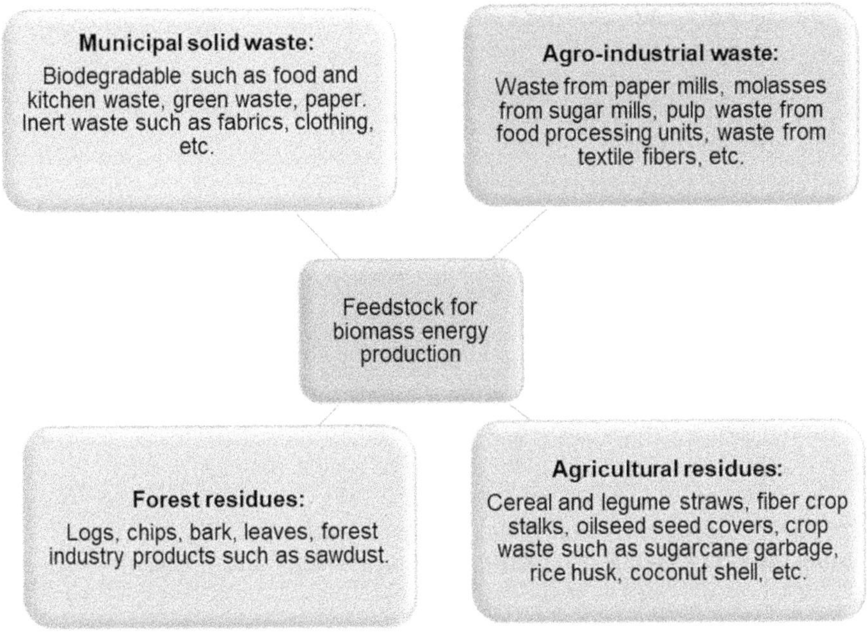

FIGURE 7.1 Raw materials for biofuel utilization.

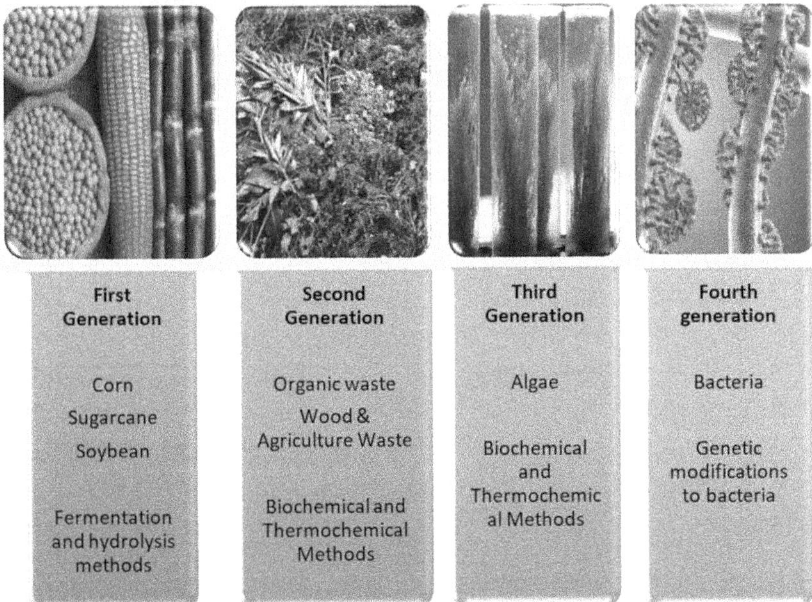

FIGURE 7.2 Different generations of biofuels.

biofuels, do not come from food sources but are produced by algae. Finally, fourth-generation biofuels use the development of cyanobacteria.[5]

7.1.2 SUSTAINABLE HARVESTING AND UTILIZATION

The production of biofuels is the subject of controversy. The climatic impact, socio-economic factors, and the production of first- and second-generation biofuels have substantially affected the rights of workers and indigenous communities due to food prices. This means that part of the land that was previously used for growing food is now being employed to grow crops for biofuel production.[8]

The consequences in the crop areas devoted to the production of biomass are in the form of large-scale monocultures, a practice that, in the medium term, depletes the soil of nutrients, making it unproductive and leading to desertification problems. The use of fertilizers, pesticides, fungicides, and large-scale irrigation systems, alone or in combination, contributes to the scarcity of good-quality soil or water.

Deforestation and organic matter residues can be used to prepare second-generation biofuels, which are often employed as natural fertilizers to nourish and restore the soil. These are other direct consequences of biofuel production.[9]

7.2 BIOFUEL PRODUCTION PROCESSES

Biomass is an abundant carbon source from which biogasoline is produced and has advantageous features over conventionally produced gasoline such as being renewable and sustainable, which, in the opinion of many authors, makes it of great market interest in the short term.[10]

Biofuel production processes involve several common steps and may vary depending on the type of biofuel being produced (biodiesel, bioethanol, biogas, etc.), feedstock (first-, second-, third-, and fourth-generation biomass), and specific employed technology.[11]

Given the diverse physical and chemical characteristics of different types of biomass, it is crucial to homogenize the raw materials as much as possible. Biomass pretreatment technologies play a vital role to this end as they encompass all intermediate process steps that modify the physical or chemical properties of a biomass resource before its use.[12]

This section describes some of the pretreatment technologies for different types of biomass, such as thermal conversion, pyrolysis and gasification, and biochemical, such as fermentation.

7.2.1 THERMOCHEMICAL CONVERSION (PYROLYSIS AND GASIFICATION)

Gasification

One route to produce biogasoline from biomass is gasification. The synthesis gas can be used for methanol production, and then, methanol is transformed into biogasoline.[13]

In gasification, biomass conversion is carried out under stoichiometric oxygen conditions, creating a reducing atmosphere that guarantees the partial oxidation of the material to be gasified and high temperatures (800–1800 °C)[3] depending on the material characteristics and the type of gasifier used. These gases can be converted into liquid fuels by the FT process.[14]

The most common sources for biosyngas production are lignocellulosic materials and microalgae because of their high thermal conversion to hydrogen and carbon monoxide (CO). However, these are not the only components of biosyngas; carbon dioxide, methane, and high-molecular-weight hydrocarbons are commonly found in biomass-derived syngas, requiring an additional purification step before introduction into any catalytic reactor.[11]

The conversion of biomass into synthesis gas through the pyrolysis and gasification processes occurs according to the main steps indicated by the following reactions:[15]

$$\text{Biomass}\left(C_xH_yO_z\right)_n \xrightarrow{\Delta} \text{gas} + \text{tars} + \text{carbon} \tag{7.1}$$

$$\text{tars} \rightarrow C_{4-7} + C_{9+} + CO + CO_2 + H_2 \tag{7.2}$$

$$C_{9+} \rightarrow CH_4 + H_2 \tag{7.3}$$

$$\text{Carbon} \rightarrow CO + CO_2 + H_2 + \text{Residual solids} \tag{7.4}$$

The reaction is a heterogeneous system that involves drying, removal, and oxidation of volatile materials from the fuel and coal, resulting in changes in particle size and pressure losses in the bed.[13]

Different types of gasifiers are classified according to their gas–solid contact mode and gasification medium[16] Based on the gas–solid contact mode, gasifiers are divided into three types: fixed or moving bed, fluidized bed, and entrained flow. Each is further subdivided into specific types as shown in Figure 7.3.[17]

Pyrolysis

Pyrolysis is the conversion of biomass into liquid (bio-oil), solid, and gaseous fractions. The process is carried out by heating the biomass in the absence of air at temperatures from around 300 to 700 °C or by using catalysts.[18]

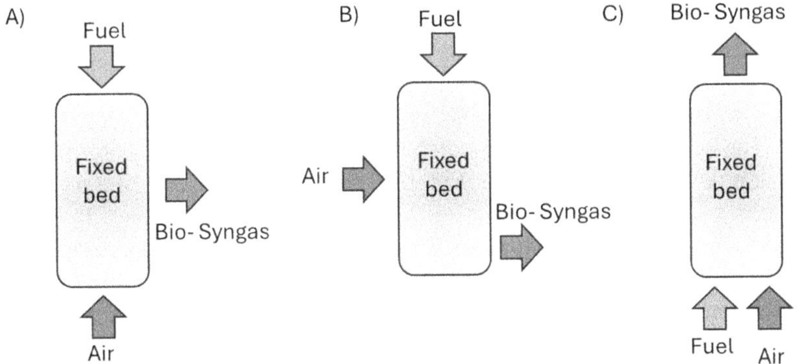

FIGURE 7.3   Types of gasifiers. (a) Fixed bed counterflow; (b) fixed bed co-current flow; and (c) circulating fluidized bed.

The reactions involved in pyrolysis are the following:[9]

$$\left(C_x H_y O_z\right)_n \xrightarrow{\Delta} nCH_4 + 3nH_2O + 2nCO + 3\,Carbon \tag{7.5}$$

$$\left(C_x H_y O_z\right)_n \xrightarrow{\Delta} 2.5nCH_4 + 1.5nH_2O + 2nCO \tag{7.6}$$

$$\left(C_x H_y O_z\right)_n \xrightarrow{\Delta} 2.5nCH_4 + 2.5nCO_2 + nC \tag{7.7}$$

$$\left(C_x H_y O_z\right)_n \xrightarrow{\Delta} 5nH_2 + 1.5nCO_2 + 2nCO + 2.5nC \tag{7.8}$$

Many types of vegetable oils have been investigated as renewable biomass-derived feedstocks for biofuel production; some of these renewable feedstocks can be classified as edible or inedible for human consumption[15] and their use depends on the geographical location of the oil source, such as jatropha, karanja, flaxseed, rubber seed, mahua, neem, and castor oil.

Bio-oil, a product of biomass pyrolysis, is challenging, as it is highly viscous, corrosive, and unstable in nature, with low calorific values and poor fuel qualities. It is a complex mixture of over 400 organic compounds, including oxygenated compounds like acids, aldehydes, ketones, alcohols, esters, ethers, glycols, and phenols. This complexity makes it unsuitable for direct use in stationary combustion engines, necessitating additional treatment. The urgency and importance of finding solutions to these challenges cannot be overlooked.[15]

The pyrolysis process can also be carried out with catalysts (catalytic cracking), which can direct the decomposition of the fatty acids present in the biomass to specific products, by reducing the temperature, and by accelerating the reaction time.[19]

7.3.2 Biochemical Conversion (Fermentation)

The carbohydrate fraction in biomass, a crucial component, can be effectively utilized by separating lignin from the cellulose and hemicellulose. This separation allows the carbohydrate part to be broken down into simple sugars, for example, by fermentation.

Fermentation, a widely used commercial process, plays a crucial role in ethanol production from crops like sugarcane, sugar beet, and starch crops such as corn and wheat. The biomass is milled, and starch is converted into sugars by enzymes. These sugars are then transformed into ethanol by yeast, a key player in the fermentation process.[20]

Ethanol purification by distillation is an energy-intensive step: approximately 450 L of ethanol is produced per every 1,000 kg of dry corn. The solid residues obtained from this process, versatile in their applications, can be fed to livestock and the bagasse obtained from sugarcane can be used for further gasification or as boiler fuel.

The use of gaseous fermentation to produce biofuels, compared in mass and energy conversion efficiencies to enzymatic lignocellulose hydrolysis, offers a number of advantages, such as the low need for pretreatment and energy for bioconversion.

However, the technology is very limited by the low productivity of ethanol, and its selling price using this technology is higher than that of gasoline.[21]

7.3 KEY BIOMASS-TO-GASOLINE TECHNOLOGIES

7.3.1 FISCHER–TROPSCH SYNTHESIS

Fischer–Tropsch (FT) synthesis, also known as indirect coal liquefaction, is a process involving the catalytic hydrogenation of CO to produce a wide variety of hydrocarbons, such as gasoline, jet fuel, diesel, and waxes. It has been increasingly incorporated as a sustainable process to produce high-performance fuels from bio-syngas. It has gained increasing interest in the industry for generating clean fuels, which can help meet the growing global energy demand and comply with future environmental regulations.[7]

In the bio-Fischer-Tropsch (BFT) process, the biomass is first converted into bio-syngas by gasification and then a cleaning process is applied to remove impurities in order to meet the requirements of FT synthesis (Table 7.1). This clean biogas is fed into a catalytic reactor to produce green gasoline, diesel, and other clean biofuels.[22]

In general, bio-syngas contains organic impurities, such as tars, benzene, toluene, and xylenes (BTX), inorganic impurities, like O_2, NH_3, HCN, H_2S, COS, and HCl, as well as dust and soot.[23]

The biggest drawback with organic impurities (aromatic compounds, oxygenated hydrocarbons, and polycyclic aromatic hydrocarbons) is that they can poison the catalyst and stop the BFT reaction.

Bio-syngas generally contains 0.5–1 vol% oxygen, nitrogen species such as NH_3, HCN, and NO_x, H_2S, dust, soot, and other impurities; various catalysts such as Cu/Zn/Al/ HZSM-5, Pd/Al_2O_3, and zirconia-yttria tubular membranes are used for their removal. Specifically for ammonia removal, aqueous scrubbers are used. NO_x can be removed by platinum and metals (Cu, Cr, and Fe) based on zeolite catalysts (H-ZSM-5).[14]

Currently, the FT process plays a key role in gas-to-liquid (GTL) technology. The reactions in this process are usually described as follows:[14]

TABLE 7.1

Clean Syn-gas Requirements for Fischer–Tropsch Synthesis

Impurity	Specification
Sulfur compounds	<1ppm[a]
Nitrogen compounds	<1 ppm
Inorganic acids	<10 ppb[b]
Heteroatoms	<1 ppm

[a] Parts per million
[b] Parts per billion

$$(2n + 1)H_2 + nCO \rightarrow C_nH_{2n+2} + nH_2O \tag{7.9}$$

$$2nH_2 + nCO \rightarrow C_nH_{2n} + nH_2O \tag{7.10}$$

$$2nH_2 + nCO \rightarrow C_nH_{2n} + 2O + (n-1)\,H_2O \tag{7.11}$$

$$CO + H_2O \leftrightarrow CO_2 + H_2 \tag{7.12}$$

In addition to alkanes (Equation 7.9) and alkenes (Equation 7.10), some oxygenates (Equation 7.11) can also be formed during the FT process. A water gas shift reaction (Equation 7.12) that occurs during the process can be used to adjust the ratio of CO and hydrogen.[24]

The chemical reaction between hydrogen and CO requires a large amount of energy to occur. Therefore, catalysts are crucial to ensure chemical conversion in appreciable amounts of products. The chemical composition of the products obtained from the FT synthesis depends largely on the catalyst selectivity and reaction extent related to the process thermodynamics.[14,25]

7.3.2 HYDROPROCESSING

Catalytic hydroprocessing of oils derived from petroleum or biomass involves hydrotreating and hydrocracking technologies to remove impurities (sulfur, nitrogen, and oxygen) from these raw materials.

Catalytic hydroprocessing of liquid biomass is a practical technology that can convert crude vegetable oils, used cooking oils, animal fats, and algal oils into biofuels with significantly high conversion yields. The catalytic process allows the transformation of triglycerides and lipids into kerosenes and isoparaffins, obtaining products in the naphtha, kerosene, and diesel ranges. These biofuels, compared to their fossil and conventional biofuel equivalents, have high calorific value, high cetane number, excellent oxidation stability, practically zero acidity, and higher saturation level. This technology is not only efficient but also practical, making it a reliable solution for improving the intermediate products of solid biomass (pyrolysis and FT oils).[7]

The catalytic hydrotreatment of liquid biomass resembles conventional oil stream procedures. A standard catalytic hydrotreatment unit comprises four fundamental sections: feed preparation, reaction, product separation, and fractionation. The hydrotreating catalytic reactions of liquid biomass depend on the type of biomass being processed, the operating conditions, and the employed catalyst as described below:[23]

a. Cracking reactions in biomass convert oils/fats such as triglycerides into their constituent fatty acids (carboxylic acids) and propane (Figure 7.4). Molecules such as triglycerides with a boiling point above 600°C are transformed into middle distillate products (naphtha, kerosene, and gas oil). Another cracking reaction that can occur in the catalytic hydrotreating of pyrolysis oil is that of aromatic and polyaromatic compounds.

b. Saturation: Excess hydrogen in the feed to the hydrotreating reactor causes the cleavage of carbon–carbon double bonds and saturation of unsaturated carboxylic acids to form saturated ones. This reaction also forms naphthenes through the conversion of unsaturated cyclic compounds and aromatic compounds (Figure 7.5).

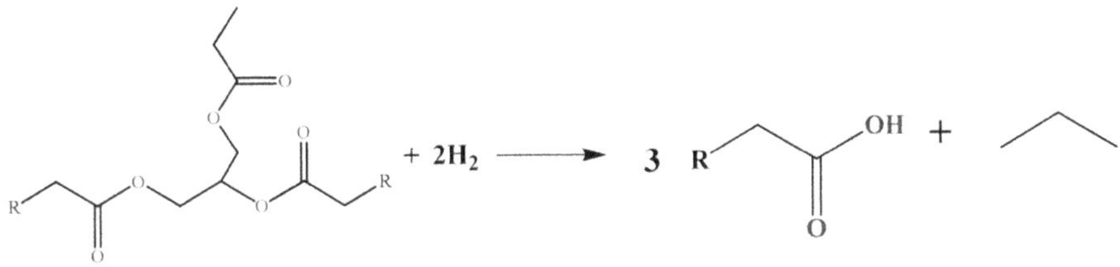

FIGURE 7.4   Triglyceride cracking.

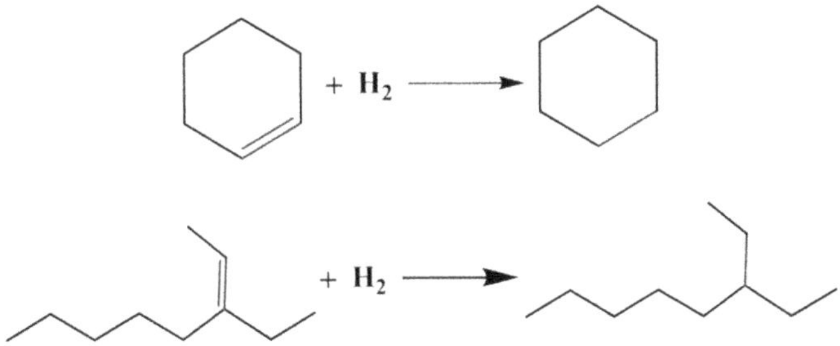

FIGURE 7.5   Saturation reaction.

c. Elimination of heteroatoms: Sulfur (S), nitrogen (N), and oxygen (O). The main reactions are deoxygenation, decarbonylation, and decarboxylation and the products are n-paraffins, water, CO, and carbon dioxide. These reactions give the paraffinic character to the produced biofuels, and for this reason, the hydrotreated products are often referred to as paraffinic fuels.

d. Isomerization: Although the straight-chain paraffinic molecules resulting from the reactions mentioned in c) increase the cetane number, calorific value, and oxidation stability in biofuels containing them, they also degrade their cold flow properties. To improve their cold flow properties, isomerization reactions are required, which usually take place during the second step/reactor as they require a different catalyst.[8]

7.4   ZEOLITE CATALYSIS

Catalytic hydroprocessing involves using catalysts and controlled temperature and pressure conditions in the presence of hydrogen to modify the chemical composition of the bio-oil. During this process, the oxygenated compounds present in the bio-oil undergo hydrogenation, hydrodeoxygenation, and dehydration reactions, removing oxygen and forming simpler, more stable hydrocarbons.

Catalysts used in catalytic hydroprocessing can vary depending on the specific conditions and bio-oil treatment objectives. Typical catalysts include sulfides of metals, such as nickel, cobalt, molybdenum, and iron, supported on alumina, silica, or zeolites. These catalysts facilitate hydrogenation and deoxygenation reactions, thus helping improve bio-oil quality.[18]

Catalytic hydroprocessing of biomass is a growing topic with much room for improvement and optimization. There are not many commercial catalysts designed and developed specifically for biomass. So, conventional commercial catalysts, such as cobalt and molybdenum (CoMo) or nickel and molybdenum (NiMo) based catalysts on alumina (Al_2O_3) substrate are used. However, the selection of the hydrotreating catalyst for biomass applications is incredibly crucial and challenging for two reasons: (a) catalyst activity varies significantly since commercial catalysts are designed for different feedstocks, that is feedstocks with high sulfur concentration, heavy feedstocks (containing large molecules), feedstocks with high oxygen concentration, etc.; b) commercial hydroprocessing catalysts are currently not available for lipid feedstocks and other intermediate products of biomass conversion processes (e.g., pyrolysis bio-oil); so, it is necessary to explore and evaluate commercial hydrotreating catalysts, as each catalyst has different performance and degradation rate.

After catalytic hydroprocessing, the treated bio-oil generally has a lower viscosity, is less corrosive, more stable, and has a higher calorific value. In addition, the content of oxygenated compounds is significantly reduced, making it more suitable to be used as a fuel in stationary combustion engines or other industrial applications.[18]

However, significant efforts have been made to develop special hydrotreating catalysts to convert/upgrade liquid biomass into biofuels.

7.5 NON-PETROLEUM FEEDSTOCKS

7.5.1 ALGAE-BASED BIOFUELS

One limitation of terrestrial biomass is its reliance on cultivating plants for energy, often at the expense of their use as a food source. In contrast, biomass derived from aquatic and marine species, particularly algae, represents an intriguing alternative due to its unique biological characteristics. Algal biomass, comprising proteins, lipids, and carbohydrates that make up to 90% of its total dry weight, also contains nucleic acids, minerals, and other components that represent the remaining 10%, offering a rich and diverse resource.[26]

Algae, fascinating eukaryotic organisms, contain chlorophyll and perform photosynthesis using oxygen, releasing water as an electron donor.[27] Unlike their terrestrial counterparts, they lack roots and conductive tissues, making them a unique and intriguing species. They are further classified as microalgae, such as diatoms and dinoflagellates, and macroalgae, such as sea lettuce.

Algae, a potential powerhouse, offers diverse biofuels, including biodiesel, butanol, gasoline, methane, ethanol, vegetable oils, and kerosene. Figure 7.6 illustrates the process of obtaining these biofuels. Importantly, these biofuels boast remarkably high yields and are considered as third-generation, underscoring their immense potential as a sustainable and efficient energy source.

Biofuels produced from algae may need to be refined through distillation, filtration, and other separation processes. They can then be used directly or blended with conventional fuels for use in vehicles, power generation, or other applications.

It is important to remember that each stage of the algal biofuel production process may require specific technologies and face technical and economic challenges that must be addressed to achieve efficient and cost-effective production.

Several microalgae species, such as *Dunaliella tertiolecta*, *Streptomyces* platensis, *Chlorella protothecoides*, *Nannochloropsis* sp., and heterotrophic *Chlorella*

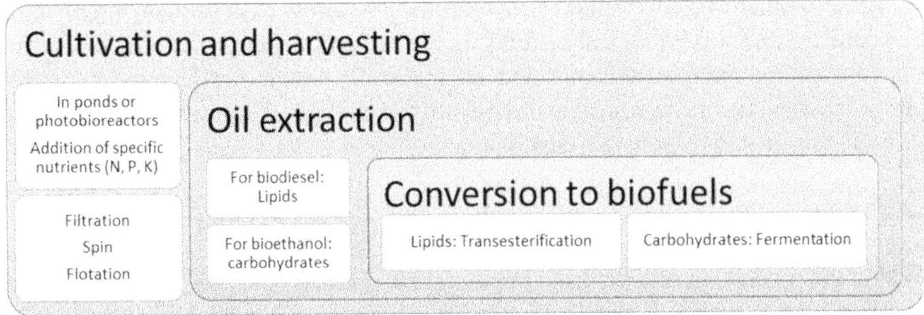

FIGURE 7.6 Processes to obtain biofuels from algae.

protothecoides, have demonstrated their potential in gasoline production. This promising development is further bolstered by the efforts of companies like Sapphire Energy, which has pioneered advanced algae farming technology with the ambitious goal of producing 100 million gallons of gasoline annually. Similarly, Exxon Mobil, in collaboration with Synthetic Genomics, Inc., is on track to produce hydrocarbon biofuels from algae in the foreseeable future, within the next 5–10 years.[28]

7.6 SYNTHETIC BIOLOGY APPROACHES

Synthetic biology is a biotechnology field focusing on designing and constructing new or modified biological systems for practical applications. Synthetic biology is used in biofuels to develop organisms specifically designed to produce biofuels more efficiently and economically. Some of the synthetic biology approaches used to produce biofuels are shown in Figure 7.7.

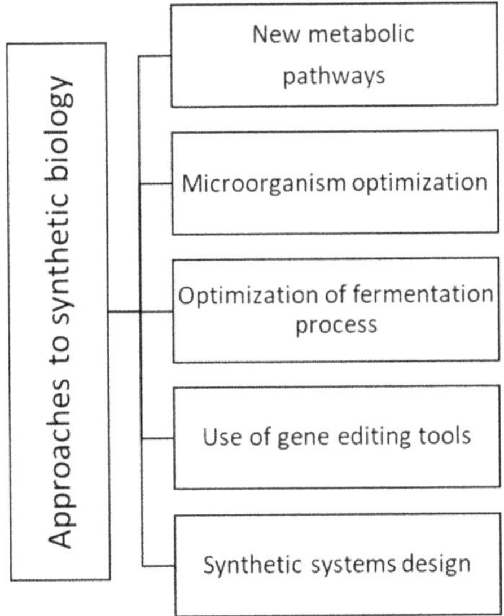

FIGURE 7.7 Synthetic biology approaches.

These synthetic biology approaches offer the possibility of developing highly efficient and customized biological systems for biofuel production, which could contribute to creating more sustainable and economically viable production processes. However, they also raise ethical and regulatory challenges that must be addressed to ensure the safe and responsible development of these technologies.

7.7 ENVIRONMENTAL BENEFITS AND CHALLENGES

Within the production of fuels from large-scale biomass production, there are significant challenges, such as depending on the scope of resource availability, time frame, and technical, economic, or environmental factors that limit its use. For example, cultivation for biomass purposes may require large amounts of water, soil nutrients, and other resources, which could generate environmental concerns and conflicts with other land uses. Other areas of opportunity to be improved are production costs and the vast space for storage. These challenges have been gradually dissipated through continuous efforts from academics, technologists, scientists, and governments because there are more benefits to be found in the use of biomass.[5]

One of the most important benefits is the reduction of GHG emissions because although the combustion of biomass emits carbon dioxide (CO_2), the carbon cycle associated with biomass is considered as carbon neutral in the long term since plants absorb CO_2 from the atmosphere during their growth. It has minimum sulfur content (amounts below 1%), which avoids the formation of acid rain, and also allows the manufacture of biodegradable products. Another benefit is that the use of advanced technologies, such as gasification and pyrolysis, can significantly reduce GHG emissions compared to fossil fuels.

The economic benefits include providing development in rural areas, either through the harvesting of energy crops, sustainable forest management, or the operation of bioenergy plants, where organic waste such as agricultural and forestry residues can be used to obtain energy; however, as mentioned before, there are still some logistical and infrastructure challenges like the collection, transportation, and processing of biomass, which require significant investment in infrastructure, especially in regions where biomass is dispersed or far from consumption centers.

7.8 CURRENT STATE OF THE BIOFUEL MARKET AND ECONOMIC PROSPECTS FOR BIOMASS-DERIVED GASOLINE

Changing environmental regulations and increasing demand for green energy sources by the growing population are driving the development of the biofuels market; consequently, research and development in this field continue advancing, making biofuels more efficient and cost-effective.

Currently, biofuels are consumed in industries such as the automotive, marine, and rail, however, several strategic initiatives are being taken to increase the global presence of biofuels, with key market developments like new product launchings. Some of these are presented in Figure 7.8.[23]

The economic viability of biomass-derived gasoline is significantly influenced by the demand for biomass. This demand, along with factors such as the cost of biomass

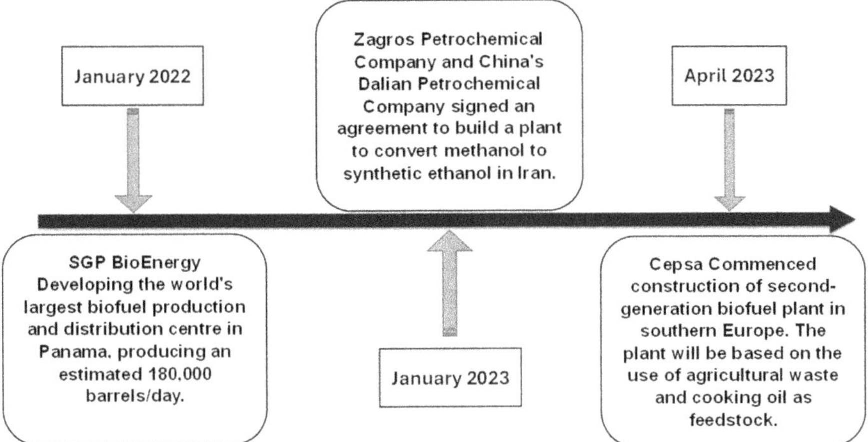

FIGURE 7.8 Investment and financing of alternative fuel technologies.

feedstock, efficiency of conversion technologies, government policies and incentives, and the market's appetite for renewable fuels, plays a crucial role in determining the profitability of the final product.

The economic viability of biomass-derived gasoline is directly linked to the availability of feedstocks for biomass production. This availability, which can be influenced by factors such as geographic location, seasonality, and competition with other uses (e.g., food production), can be optimized by choosing cheaper feedstocks and large-scale production technology, thereby improving the overall viability of gasoline production from biomass.

One of the most substantial challenges faced for its use is the access to markets and the ability to compete with petroleum-derived gasoline prices, all of which are important considerations for the economic viability of biomass-to-biogasoline projects. In general, while biomass-derived gasoline holds promise as a renewable alternative to fossil gasoline, its economic prospects depend on various factors and may vary depending on specific circumstances and market conditions. Notwithstanding, it is important to note that continued technological innovation, supportive policies, and market development efforts are not just beneficial, but essential to developing the economic potential of biomass-derived gasoline as part of the transition to a more sustainable energy future.

7.9 REGULATORY LANDSCAPE

Governments all over the world have implemented various incentives and policies to improve the use of biomass for biofuel production. These initiatives are part of a global strategy to reduce GHG emissions and promote the development of clean and renewable technologies. One of these efforts is to provide tax credits, direct subsidies, or financial incentives to help offset biofuel production costs and promote investment in biomass and biofuel production facilities. Carbon credits have also been provided, allowing companies to generate credits by reducing GHG emissions. The production and use of biofuels can generate carbon credits due to their ability to reduce emissions compared to fossil fuels. These credits can be traded on carbon markets and provide an additional source of revenue for biofuel producers.

Another incentive mechanism is supporting mechanisms that guarantee a fixed price for energy produced from renewable sources, including biomass. These guaranteed prices not only make biofuel production more profitable but also make it an attractive investment opportunity. This financial boost encourages investment in biomass-to-biofuel conversion technologies, promising a bright future for the biofuel industry.

These are just a few examples of government policies and incentives used to promote the use of biomass for biofuel production. The beauty of these policies is their adaptability. They vary by country and may evolve in response to changing political, technological, and economic priorities. This adaptability ensures the stability and growth of the biofuel industry, making it a reliable and promising sector to invest in.

7.9.1 Compliance and Certification Standards

Compliance and certification rules for the use of biomass in biofuel production vary by country and region and may be subjected to different standards and specific requirements. However, some widely recognized international standards and certification systems are used to ensure the sustainability and quality of biofuels produced from biomass. On the one hand, certification systems provide the criteria and requirements to ensure specific environmental, social, and economic standards for biogasoline production and quality standards set for the technical and performance requirements for biofuels, ensuring that they meet the quality and safety standards for their use in specific applications, such as transport or power generation. Chain-of-custody certification systems ensure the traceability and tracking of biomass and biofuels throughout the supply chain, from feedstock to final product. This helps ensure that the produced biofuels meet the established sustainability and quality standards. Some examples of chain-of-custody certifications are shown in Figure 7.9.

Biomass and Bio-gasoline Standards	Sustainability certification:
	*RSB (Roundtable on Sustainable Biomaterials)
	*ISCC (International Sustainability and Carbon Certification)
	*RSPO (Roundtable on Sustainable Palm Oil) for biofuels derived from palm oil
	*Bonsucro for biofuels derived from sugarcane bagasse
	*Bonsucro for biofuels derived from sugarcane bagasse
	*RSPO (Roundtable on Sustainable Palm Oil) for biofuels derived from sugarcane bagasse
	Quality standards:
	*ASTM D6751 for biodiesel
	*EN 14214 for biodiesel in Europe
	*ASTM D4806 for fuel-grade ethanol
	*EN 15376 for fuel-grade ethanol in Europe
	Chain of Custody certification
	*RED cert
	*ISCC PLUS
	*RSB EU RED

FIGURE 7.9 Certifications and standards for the use of biomass in fuel production.

It is important to note that certification and compliance may vary depending on the type of biomass used, production process, biofuel destination, and specific regulations of each country or region. Companies seeking to certify their biofuel products typically undergo independent audits to verify compliance with the standards and requirements set by the relevant certification schemes.

7.9.2 GLOBAL INITIATIVES PROMOTING NON-PETROLEUM FUELS

Some of the global initiatives that promote the use of nonpetroleum fuels as part of the efforts to address climatic change, reduce dependence on fossil fuels, and promote energy sustainability are:

The Paris Agreement: Aims at limiting global warming below 2 degrees Celsius, above pre-industrial levels, and strives to limit it to 1.5 degrees Celsius. The agreement recognizes the importance of reducing GHG emissions and promotes the use of renewable and low-carbon energy sources, including biofuels and other nonpetroleum fuels.

Sustainable Development Goals (SDGs): Goal 7 of the SDGs, adopted by the United Nations General Assembly in 2015, seeks to ensure access to affordable, reliable, sustainable, and modern energy for everybody as well as to increase the share of renewable energies in the global energy matrix. The promotion of nonpetroleum fuels is in line with these objectives.

Decarbonization initiatives: Several countries and regions have launched initiatives to decarbonize their economies and reduce carbon emissions. These initiatives often include policies and programs to encourage the adoption of renewable energy sources and the transition to a cleaner energy mix, which may include the use of biofuels, green hydrogen, and other nonpetroleum fuels.

Clean Energy Partnership: This is a global initiative that tries to accelerate the transition to clean and sustainable energy worldwide. The alliance brings together governments, companies, investors, and other stakeholders to promote investment in renewable energy, improve energy efficiency, and foster innovation in clean technologies, including nonpetroleum fuels.

Regional and national initiatives: Many countries and regions have their initiatives and policies to promote the use of biofuels, renewable electricity, green hydrogen, and other forms of nonpetroleum energy. These initiatives may include renewable fuel standards, tax incentives, research and development support programs, and renewable energy targets.

7.10 CHALLENGES AND FUTURE OUTLOOK

The modern world is facing numerous challenges, such as energy security, oil prices, resource depletion, and climatic change. All of these are directly or indirectly damaging the environment. All these challenges have led to remarkable advances in research and development of renewable energy and fuels, among which the use of biomass stands out.

Undoubtedly, biofuels is a fast-growing and fast-moving research field. Significant progress has been made in biofuel research, but fossil fuels cannot be completely replaced yet. Several integrated engineering and biological approaches are being applied to optimize biofuel production on a commercial scale.

7.11 CONCLUSION

In conclusion, biomass presents several significant advantages regarding its potential to provide renewable energy and mitigate the environmental impacts of fossil fuel use. However, it also poses significant challenges that must be addressed to ensure its positive contribution to the transition to a more sustainable and resilient energy system.

The estimated amount of fossil fuels burned over a year is equivalent to more than 400 times the productivity of the planet's current plant life. This means that the amount of material used annually is equivalent to the number of plant species that have developed on the planet over four centuries. This level of consumption implies that biofuels cannot replace all the fossil fuels currently in use, so research and technological development must continue to achieve sustainable energy projects.

REFERENCES

1. Kumar A, Kuang Y, Liang Z, Sun X. Microwave chemistry, recent advancements, and eco-friendly microwave-assisted synthesis of nanoarchitectures and their applications: a review. *Mater Today Nano.* 2020;11:100076. doi:10.1016/j.mtnano.2020.100076

2. Montoya Arbeláez JI, Chejne Janna F, Garcia-Pérez M. Fast pyrolysis of biomass: A review of relevant aspects. Part I: Parametric study. *DYNA.* 2015;82(192):239–248. doi:10.15446/dyna.v82n192.44701

3. Kumar A, Kumar N, Baredar P, Shukla A. A review on biomass energy resources, potential, conversion and policy in India. *Renew Sustain Energy Rev.* 2015;45:530–539. doi:10.1016/j.rser.2015.02.007

4. Huber GW, Iborra S, Corma A. Synthesis of transportation fuels from biomass: chemistry, catalysts, and engineering. *Chem Rev.* 2006;106(9):4044–4098. doi:10.1021/cr068360d

5. Malode SJ, Prabhu KK, Mascarenhas RJ, Shetti NP, Aminabhavi TM. Recent advances and viability in biofuel production. *Energy Convers Manag X.* 2021;10:100070. doi:10.1016/j.ecmx.2020.100070

6. Stephen JL, Periyasamy B. Innovative developments in biofuels production from organic waste materials: A review. *Fuel.* 2018;214:623–633. doi:10.1016/j.fuel.2017.11.042

7. Rodionova MV, Poudyal RS, Tiwari I, et al. Biofuel production: Challenges and opportunities. *Int J Hydrog Energy.* 2017;42(12):8450–8461. doi:10.1016/j.ijhydene.2016.11.125

8. Fang Z, ed. *Liquid, Gaseous and Solid Biofuels - Conversion Techniques.* InTech; 2013. doi:10.5772/50479

9. Romanelli GP, Ruiz DM, Pasquale GA. *Química de la biomasa y los biocombustibles.* Editorial de la Universidad Nacional de La Plata (EDULP); 2016. doi:10.35537/10915/59392

10. Bakhtyari A, Makarem MA, Rahimpour MR. Light olefins/bio-gasoline production from biomass. In: *Bioenergy Systems for the Future.* Elsevier; 2017:87–148. doi:10.1016/B978-0-08-101031-0.00004-1

11. Rauch R, Hrbek J, Hofbauer H. Biomass Gasification for Synthesis Gas Production and Applications of the Syngas. In: Lund PD, Byrne J, Berndes G, Vasalos IA, eds. *Advances in Bioenergy.* 1st ed. Wiley; 2016:73–91. doi:10.1002/9781118957844.ch7

12. Rezania S, Oryani B, Cho J, et al. Different pretreatment technologies of lignocel-lulosic biomass for bioethanol production: An overview. *Energy.* 2020;199:117457. doi:10.1016/j.energy.2020.117457

13. Pérez J, Borge D, Agudelo J. Proceso de gasificación de biomasa: una revisión de estu-dios teórico – experimentales. *N° 52* pp 95–107. Published online 2010.

14. Hu J, Yu F, Lu Y. Application of Fischer–Tropsch Synthesis in Biomass to Liquid Conversion. *Catalysts.* 2012;2(2):303–326. doi:10.3390/catal2020303

15. Alipour Moghadam R, Yusup S, Azlina W, Nehzati S, Tavasoli A. Investigation on syngas production via biomass conversion through the integration of pyrolysis and air–steam gas-ification processes. *Energy Convers Manag.* 2014;87:670–675. doi:10.1016/j.enconman.2014.07.065

16. Mishra S, Upadhyay RK. Review on biomass gasification: Gasifiers, gasifying mediums, and operational parameters. *Mater Sci Energy Technol.* 2021;4:329–340. doi:10.1016/j.mset.2021.08.009

17. Basu P. Design of Biomass Gasifiers. In: *Biomass Gasification Design Handbook.* Elsevier; 2010:167–228. doi:10.1016/B978-0-12-374988-8.00006-4

18. Chiaramonti D, Prussi M, Buffi M, Rizzo AM, Pari L. Review and experimental study on pyrolysis and hydrothermal liquefaction of microalgae for biofuel production. *Appl Energy.* 2017;185:963–972. doi:10.1016/j.apenergy.2015.12.001

19. Haryani N, Harahap H, Taslim I. Biogasoline production via catalytic cracking process using zeolite and zeolite catalyst modified with metals: a review. *IOP Conf Ser Mater Sci Eng.* 2020;801(1):012051. doi:10.1088/1757-899X/801/1/012051

20. Ullah K, Sharma VK, Ahmad M, et al. The insight views of advanced technologies and its application in bio-origin fuel synthesis from lignocellulose biomasses waste, a review. *Renew Sustain Energy Rev.* 2018;82:3992–4008. doi:10.1016/j.rser.2017.10.074

21. Bustamante García V, Carrillo Parra A, Prieto Ruíz JÁ, Corral-Rivas JJ, Hernández Díaz JC. Química de la biomasa vegetal y su efecto en el rendimiento durante la torrefacción: revisión. *Rev Mex Cienc For.* 2017;7(38):5–24. doi:10.29298/rmcf.v7i38.8

22. Manzer LE, Van Der Waal JC, Imhof P. The Industrial Playing Field for the Conversion of Biomass to Renewable Fuels and Chemicals. In: Imhof P, Van Der Waal JC, eds. *Catalytic Process Development for Renewable Materials.* 1st ed. Wiley; 2013:1–24. doi:10.1002/9783527656639.ch1

23. Singh RS, Pandey A, Gnansounou E, eds. *Biofuels: Production and Future Perspectives.* Taylor & Francis; 2017. https://catalog.loc.gov/vwebv/search?searchCode=LCCN&searchArg=2016012567&searchType=1&permalink=y

24. Millinger M, Ponitka J, Arendt O, Thrän D. Competitiveness of advanced and conven-tional biofuels: Results from least-cost modelling of biofuel competition in Germany. *Energy Policy.* 2017;107:394–402. doi:10.1016/j.enpol.2017.05.013

25. Santos RGD, Alencar AC. Biomass-derived syngas production via gasification process and its catalytic conversion into fuels by Fischer Tropsch synthesis: A review. *Int J Hydrog Energy.* 2020;45(36):18114–18132. doi:10.1016/j.ijhydene.2019.07.133

26. Vega BOA, Lobina DV. *Métodos y Herramientas Analíticas En La Evaluación de La Biomasa Microalgal.* Centro de Investigaciones Biológicas del Noroeste; 2007. https://books.google.com.mx/books?id=_0IVAQAAIAAJ

27. Wood DA. Microalgae to biodiesel - Review of recent progress. *Bioresour Technol Rep.* 2021;14:100665. doi:10.1016/j.biteb.2021.100665

28. Shamsul NS, Kamarudin SK, Rahman NA. Conversion of bio-oil to bio gasoline via pyrolysis and hydrothermal: A review. *Renew Sustain Energy Rev.* 2017;80:538–549. doi:10.1016/j.rser.2017.05.245

8 Gasoline and Fuel Conversion from Greenhouse Gases: CO_2

Heriberto Díaz Velázquez
Mexican Petroleum Institute, Mexico City, México

8.1 INTRODUCTION

Between November and December 2023, the Conference of the Parties of the UNFCCC (COP 28) took place in Dubai, where the signing countries made the compromise to leave behind the use of fossil fuels, calling it the "beginning of the end" of the fossil fuel era.[1] This means that oil and carbon-based fuels must be eliminated as soon as possible as energy sources, due to the high levels of CO_2 emissions that involve their use. It is well known that the current climate situation would not be as that we are facing if there were no human greenhouse gas (GHG) emissions as they have been since the industrial revolutions.[2] Efforts to alleviate the massive CO_2 emissions to the atmosphere have been carried out through the development of carbon capture and storage (CSS) technologies;[3–6] however, this surplus CO_2 would be only occupying empty spaces with no utility if their potential is not exploited. It is well known that CO_2 possesses very high chemical stability, which makes it not so simple to convert it into different chemicals or fuels. The concept of CO_2 conversion has been studied for several decades, and the variety of materials that have been produced with CO_2 as the raw material at the lab scale is very high;[7–11] nonetheless, its transformation through commercially available processes is scarce. Few examples are the production of urea, inorganic and organic carbonates, polycarbonates, salicylic acid, and more recently, methanol, synthesis gas (syngas) but this list can grow if novel discoveries in catalyst development are taken to the demonstration plants. In this context, it is mandatory, and urgent to innovate in the way we obtain fuels, such as gasoline, so that this can offset the net GHG emissions in order to become carbon-neutral fuels. By doing this, efforts to retro-synthesize hydrocarbons (HCs) from GHGs have been carried out in the last decade, especially by converting CO_2 into gasoline (C_8–C_{11}), aviation fuel (C_8–C_{16}), and diesel (C_{10}–C_{20}) fuel-range HCs with some success.[12,13] On the other side, there is methanol, a chemical that can be used as part of the gasoline blend itself, which is being already marketed in part as a product of CO_2 conversion.[14] It is thus of great significance to address these scientific advancements to motivate the academic and industrial communities to continue in this way. The chapter is divided into two main sections: the first one refers to the

DOI: 10.1201/9781003517283-8

conversion of CO_2 into liquid HCs by catalytic thermal hydrogenation and the second one will deal with the synthesis of methanol from CO_2 by the same procedure. Although many other synthetic routes for the conversion of CO_2 have been reported and studied at the laboratory level, catalytic thermal hydrogenation is one of the more promising outcomes for the industry. Other routes to produce fuels, such as the use of biomaterials and biomass, are also very important and has previously been discussed in more detail along this book.

8.2　DIRECT CONVERSION OF CO_2 INTO FUELS

8.2.1　REDUCTIVE HYDROGENATION OF CO_2 INTO HYDROCARBONS

Most of the investigations related to the transformation of CO_2 into fuels have been evoked to the production of methanol, either by catalytic hydrogenation, electrocatalytic, photo-electrocatalytic, or biocatalytic reactions.[13,14] On the other hand, much fewer works on the direct synthesis of large HCs are currently in the literature. The synthesis of HCs from CO_2 can be carried out by the direct or indirect way; the indirect way goes through the transformation of methanol as an "intermediate" raw material for gasoline production.[15–17] Conversely, CO_2 can also be directly hydrogenated to C_5+ by multifunctional catalysts, which is a less-efficient process but yields long-chain HCs that can be used directly as gasoline, diesel fuel, aviation fuel, or non-transportation fuel. In this section, the direct synthesis of long-chain HCs and methanol will be discussed in some detail, with emphasis on the catalytic aspects, mechanism of reaction, and reaction process.

8.2.2　SYNTHESIS OF LONG-CHAIN HYDROCARBONS FROM CO_2

Basically, the synthesis of liquid HCs from CO_2 can be achieved by direct hydrogenation of CO_2, where, two processes are involved, the first one is the production of carbon monoxide (CO) from CO_2 through the reverse water–gas shift reaction (RWGS) (Eq. 8.1),[18,19] and the second one includes further hydrogenation of CO to achieve higher HCs through the Fischer–Tropsch synthesis (FTS) (Eq. 8.2).[20–22] The two processes must be promoted by catalytic materials that encompass in their whole structure active sites for both reactions and, at the same time, maintain high activity and selectivity toward the target products. RWGS (Eq. 8.1) is the process by which CO_2 is hydrogenated to reversibly yield CO and water as products:

$$CO_2 + H_2 \leftrightarrow CO + H_2O \quad \Delta H^0 = 41 \text{ KJ/mol} \quad (8.1)$$

This process has been long studied and it is industrially available, especially to produce hydrogen in conjunction with CO reforming of methane,[17,23] which leads to the production of syngas (H_2/CO). Consequently, when syngas is reacted under the proper conditions, it can yield long-chain saturated HCs in the FTS synthesis:

$$(2n + 1)H_2 + nCO \leftrightarrow C_nH_{2n+2} + nH_2O \quad \Delta H^0 = -166 \text{ KJ/mol} \quad (8.2)$$

These HC mixtures can fall within the gasoline range (C_5–C_{11}), aviation fuel range (C_8–C_{18}), or diesel fuel range (C_{10}–C_{20}). The higher the HC range, the more variable the application. Then, as a direct result of CO_2 hydrogenation, the overall synthesis of liquid HCs from CO_2 can be depicted in the overall reaction as follows:

$$CO_2 + H_2 \rightarrow C_nH_{2n+2} + nH_2O \quad \Delta H^0 = -125 \text{ KJ/mol} \tag{8.3}$$

In order to realize this highly desirable transformation process, a highly active promoting catalyst should be applied. At this stage, there is no direct CO_2 transformation to HC plants commercially available; however, recent research efforts are pointing to this target. In this section, we will discuss some of the most outstanding investigations on catalyst technologies, the most relevant progress on catalyst developments will be covered, as well as the specific reaction paths involved.

8.2.3   Catalysts for the Synthesis of Liquid Hydrocarbons from CO_2

The direct hydrogenation of CO_2 to produce liquid HCs has been studied since the early 2000s, with Aresta as one of the pioneers on this issue;[24] however, it has been much less studied than RWGS or FTS separately. The main catalyst active metals studied are based on Co, Fe, and In; however, these materials have their own drawbacks considering their activity and stability under the reaction conditions. Here, we will cover the basics of the Co and Fe-based catalysts. Some other recent investigations will also be mentioned.

8.2.3.1   Co-based Catalysts

Due to their high activity, low cost, and easy availability, materials based on Co have been used as catalytic materials for the direct hydrogenation of CO_2. Cobalt-based materials are known to possess high activity toward FTS, but their RWGS activity remains low. This is why the catalytic materials must be boosted by the incorporation of, for example, highly hydrogenating agents such as noble metals (Pd, Pt, and Ni). Other approximations are doping with high electron-donating metals (Na and K) or the preparation of supports that can hold the reaction intermediates toward the desired products. The use of noble metals was documented by Dorner et al.,[25] who investigated the application of the catalyst system Co-Pt/Al_2O_3 to hydrogenate CO_2. This material, however, showed poor yield toward long-chain HCs, and, instead, yielded high methane concentration as a product. On the other hand, doping Co catalysts with alkaline metals was reported by Iloy et al.,[26] by the use of K to promote the activity of Co in K-Co/SiO_2. This resulted in a slightly higher amount of liquid HCs, but still low for practical applications. The improvement in the yield of long-chain HCs was achieved by Shi et al.,[27] using a bimetallic CoCu catalyst supported on TiO_2 and doped with K, where the effect of the metal doping was evident, since the addition of K allowed the production of long-chain HCs, avoiding the production of methane; however, the elimination of K provoked an abrupt increase in methane yield (90%). The addition of K as a dopant had several implications; on one side, it can hinder the methanation sites, allowing the growth of the HC chain; on the other side, it could also inhibit the adsorption of CO_2 by the catalyst surface, significantly reducing its

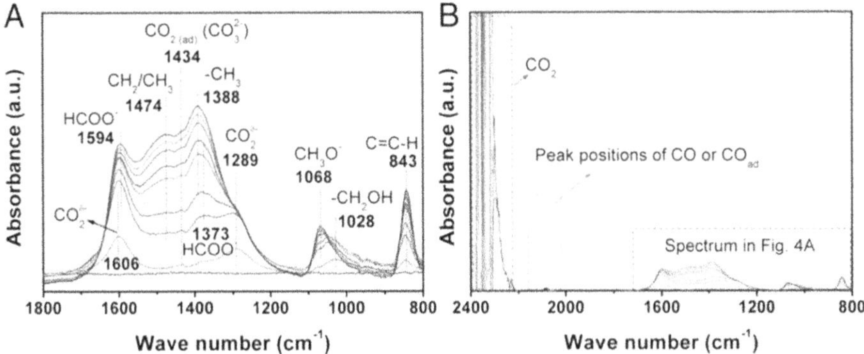

FIGURE 8.1 FT-IR spectra of the key intermediates (a) during the CO_2 adsorption on Co_6/MnO_x after reduction with H_2 at 200°C, taken at periods of time between 2 and 55 min. (b) Extended spectra. Reproduced with permission from: He, Z.; Cui, M.; Qian, Q.; Zhang, J.; Liu, H.; Han, B. Synthesis of Liquid Fuel via Direct Hydrogenation of CO_2. *Proc. Natl. Acad. Sci.* 2019, *116* (26), 12654–12659. https://doi.org/10.1073/pnas.1821231116.

conversion and increasing the selectivity to CO. These reactions were carried out at 250°C, under a $H_2/CO_2/N_2$ flow of 73:24:3 mole ratio (5 MPa). The production of liquid HCs from CO_2 was further investigated by He et al.,[28] who reported on low-temperature CO_2 hydrogenation by the nanocluster Co_6/MnO_x system. The system reached a selectivity to C_{5+} HCs higher than 50%, which was achieved thanks to a completely different mechanism, which does not involve CO as an intermediate species, opening a new route that avoids the FTS steps. Some of the intermediates found by FT-IR were CO_2^-, $HCOO^-$, -CH_2OH, and -CH_3 (Figure 8.1a); however, no free or adsorbed CO was observed (Figure 8.1b). The size and shape of the Co nanoparticles are accounted for the high efficiency of the active phase; meanwhile, Mn was responsible for the enhanced CO_2 adsorption and the reduction of molecular H_2 uptake. The reactions were carried out at 200°C, in batch experiments ($H_2/CO_2 = 1$, P = 8 MPa) using squalane as the solvent at the optimized conditions. In total, 53% selectivity of HCs between C_5–C_{26} was achieved and almost no CO was observed (0.4 C-mol%).

8.2.3.2 Iron-based Catalysts

The reports on Fe as an outstanding catalytic material for RWGS and FTS reactions led to the research on Fe-based materials as direct CO_2 hydrogenation catalysts. Important progress was made by groups such as that of Dorner et al.,[29] who sized the reactivity of Fe toward FTS and developed a material of the weight composition 17%Fe/8%K/12%Mn/Al_2O_3, which was active toward CO_2 reduction at 290°C, 1.4 MPa, and H_2/CO_2 feed gas of 3:1. The reaction mechanism involves the transformation of CO_2 into olefins, followed by oligomerization reactions to yield jet fuel–range HCs (C_8–C_{16}). With using the FeKMn catalyst, an olefin/paraffin ratio of 4.2 at a 62.4% selectivity for C_2–C_{5+} could be was obtained. The results demonstrated the active role of K and Mn, with Using the first suppressing the hydrogenation of products, but at the same time, increasing the content of olefins by the formation of $KAlH_4$ species,

TABLE 8.1

Product Outcome of Fe/Al$_2$O$_3$-based Catalysts with Different Compositions of Fe, K, and Mn[29]

Catalyst	Selectivity (% Carbon Base)			Olefin/Paraffin	CO$_2$ Conversion (%)
	C$_1$	C$_2$–C$_{5+}$	CO Yield		
Fe 9%	55.1	34.9	10.1	0.2	18.2
Fe 17%	54.9	35.1	10.0	0.2	29.2
Fe 25%	54.9	35.1	10.0	0.2	32.1
Mn 4% Fe 17%	42.0	46.5	11.5	0.7	34.4
Mn 12% Fe 17%	34.0	55.3	10.7	1.2	37.7
Mn 20% Fe 17%	40.7	43.4	15.8	0.9	25.9
K 2% Mn 12% Fe 17%	29.4	60.1	10.5	3.8	39.5
K 8% Mn 12% Fe 17%	26.0	62.4	11.5	4.2	41.4

and the second, by enhancing the conversion of CO_2 and favoring the production of olefins over methane and alkanes by metal deposition on Fe active sites and reducing the particle size of the metal centers (Table 8.1). The production of olefins is an unavoidable side reaction process in the Co and Fe-based FT process,[20–22] and, depending on the length of the olefin chains, a suitable oligomerization catalyst must be applied to secure the formation of liquid HCs; therefore, a low olefin-to-paraffin ratio is preferred; otherwise, a good oligomerization catalyst, such as acid-based catalysts, can be applied.

In a further study, Dorner et al.[30] reported on the application of multiwalled carbon nanotubes (MWCNT) as supports for the KMnFe catalyst system, achieving a lower olefin/paraffin ratio (3.1%), but with higher CH_4 selectivity (30%) and high C_2–C_{5+} selectivity (54.3%). The effect of the MWCNT on the formation of smaller particle-size metals was utterly the motivation for these results, contributing to a higher hydrogenation rate instead of the formation of unsaturated HCs. These results motivated them to find suitable dopants, such as Ce, and adding them to the FeKMn system in FeKMn/Ce/Al$_2$O$_3$. In this case, the conversion of CO_2 was increased to 50.4%, from the 41.4% obtained without the influence of Ce, although the olefin/paraffin ratio increased by 5%.

The use of Fe-based catalytic materials has been long attractive to several groups due to the high availability and affordability of Fe precursors, like the material reported by Albrecht et al.,[31] who used a cellulose-templated synthesis of Fe$_2$O$_3$ derived from Fe(NO$_3$)$_3$ without the addition of dopants and demonstrated the high versatility of this simple system in producing a similar outcome to that obtained with the aid of dopants (K, Mn, and Ce). The reactions carried out at 350°C, 1.5 MPa, and a H$_2$/CO$_2$ ratio of 3 led to a 40% conversion of CO_2 and 37% selectivity toward C_2–C_{5+} HCs. One of the properties of this catalyst was its ease of reduction (in situ) to iron carbide, the active phase for the FT reaction. Furthermore, methane formation by both direct CO_2 and CO hydrogenation was inhibited over this catalyst.

As with Co-based species, the influence of dopants significantly determines the selectivity of hydrogenated products as well as the conversion of CO_2. One such analysis was carried out by Amoyal et al.,[32] who investigated the effect of K on spinel-based Fe-Al-O and Fe_5C_2 materials. The results showed the enhancing effect of K in improving the RWGS reaction on the spinel oxide phase by one order of magnitude. Their findings suggest that the dopant atoms form oxygen vacancies due to the reductive power of K, and, as was seen before, lead to the inhibition of hydrogen adsorption, which hinders the formation of CH_4, thus enhancing the catalytic activity to higher HCs by 1.5 times.[32] Similarly, Meiri et al.[33] came to the same conclusion using similar catalysts. Figure 8.2 shows the CO_2 adsorption TPD spectra recorded for the K/Fe-Al-O catalyst. The peak at about 150°C represents a weak surface basic site. Furthermore, an increase in the basicity by two-fold was obtained with a 4% K load. Further K load creates strong basic sites (550 and 850°C), which is the result of high interactions of K with Fe-Al-O spinel sites in the form of K_2O species.

The effect of alkaline metals was further studied by Choi et al., who used $ZnFe_2O_4$ doped with Na. The combination of Zn and Na enhanced CO_2 adsorption, thus

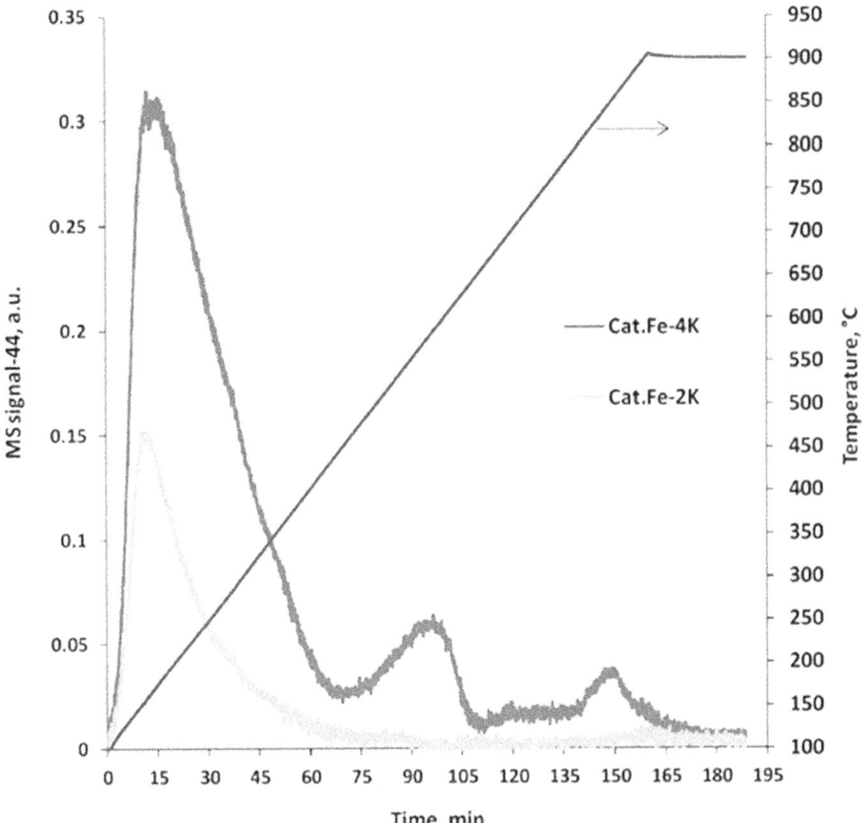

FIGURE 8.2 TCO_2 TPD spectra of K/Fe-Al-O spinel catalysts. Reproduced with permission from: Meiri, N.; Dinburg, Y.; Amoyal, M.; Koukouliev, V.; Nehemya, R. V.; Landau, M. V.; Herskowitz, M. Novel Process and Catalytic Materials for Converting CO_2 and H_2 Containing Mixtures to Liquid Fuels and Chemicals. *Faraday Discuss.* 2015, *183* (0), 197–215. https://doi.org/10.1039/C5FD00039D.

TABLE 8.2
Product Outcome of Choi's Fe-based Catalysts on CO_2 Conversion to HCs[34]

Catalyst	Selectivity			Olefin/Paraffin	CO_2 Conversion (%)
	C_1	C_2-C_{5+}	CO		
Fe_2O_3	40.5	59.5	20.5	0.16	18.2
ZnO-Fe_2O_3	53.1	46.9	23.2	0.09	24.7
$ZnFeO_4$, Na-free	43.6	56.4	21.9	0.4	27.8
$ZnFe_2O_4$, 0.08 wt% Na	9.7	90.3	11.7	11.3	34.0

provoking an abrupt increase in the selectivity toward the diesel range (about 60%), a lower CO selectivity than other Fe-based catalysts, but a higher olefin to paraffin/ ratio of 11 (Table 8.2). The alkaline metal promoted a higher dispersion of the Fe particles than that present in the non-doped catalyst. Further studies by Choi et al.[34] on delafosite-$CuFeO_2$ to produce in a single step liquid HCs from CO_2, was achieved by preventing the RWGS reaction. The material was prepared by in situ carburization of the delafossite-$CuFeO_2$ to obtain active-phase X-Fe_5C_2. The outcome yield was an excellent 66.3% yield of C_{5+} liquid HCs, but still a high olefin-to-paraffin ratio of 7.3 was obtained in their experiments.

Great progress in the production of liquid fuels at high selectivity was achieved by the group of Wei et al.,[35] who studied the catalyst system based on Na-Fe_3O_4/HZSM-5. This catalyst system is based on a three-component composite (Fe_3O_4, Fe_5C_2, and acidic zeolite) and the closeness of the two Fe phases into an associated interaction contributed to a highly efficient hydrogenation of CO_2 to the C_5–C_{11} gasoline-HC range. The great stability of the catalyst at time-on-stream experiments makes this material highly promising for its application at the industrial level (Figure 8.3a).

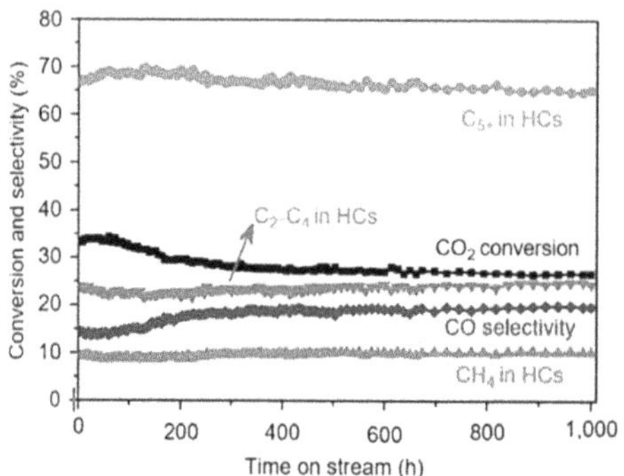

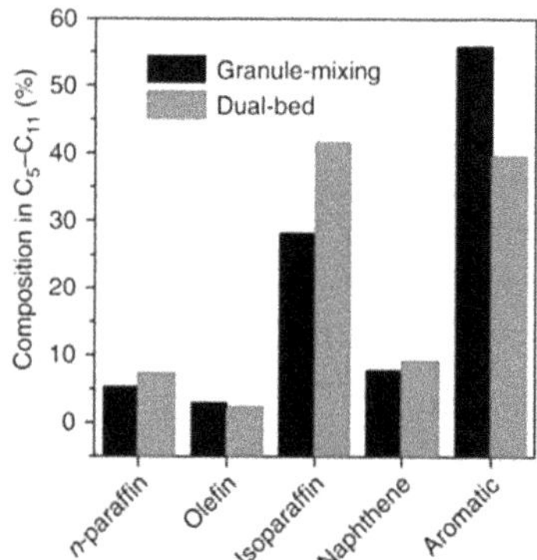

FIGURE 8.3 (a) Stability of the Na–Fe3O4/HZSM-5 catalyst with time-on-stream under the same reaction conditions. (b) Composition of gasoline-range HCs depending on the integration mode of catalyst components. Available via Creative Commons CC license by: Wei, J.; Ge, Q.; Yao, R.; Wen, Z.; Fang, C.; Guo, L.; Xu, H.; Sun, J. Directly Converting CO_2 into a Gasoline Fuel. *Nat. Commun.* 2017, *8* (1), 15174. https://doi.org/10.1038/ncomms15174.

The excellent selectivity of this catalyst toward gasoline-range HCs was evidenced by a surprising 78% selectivity to C_5–C_{11}-range HCs. The reactions were carried out at 320°C, 3 MPa, and $H_2/CO_2 = 3$ at a GHLV of 4000 mL/g cat h. Furthermore, the integration manner of this multifunctional catalyst also influenced the products distribution since naphthalene, olefins, and aromatics were also possible if granule-mixing or dual-bed mixing were used as preparation methods. This is because the acid functionality (zeolite) allows for the aromatization/isomerization/oligomerization reactions on the CO_2 hydrogenation products, allowing to obtain HCs of high-octane content within the range of paraffins, iso-paraffins, olefins, naphthenes, and aromatics (Figure 8.3b). Once more, the effect of the alkaline metal dopant was evident, causing exceptional selectivity to C_{5+} HCs and negligible methane selectivity (4%) at low H_2/CO_2 ratio, thus reducing the costs that involved the use of high hydrogen concentrations.

Copper has long been studied as the catalyst's active phase for the reduction of CO_2 to methanol, for which it has its highest selectivity.[13,36–38] The synthesis of methanol from CO_2 will be discussed as a part of this chapter. In the investigation to understand the effect of Cu on the Fe-based catalysts, Liu et al.[39] made a good contribution. The addition of Cu enhanced the production of olefins, by allowing higher adsorption of CO_2 through Cu–Fe interaction. This led to the oligomerization reactions that, ended with the conversion of C_2–C_4 olefins to their saturated counterparts and C_{5+} paraffins (Figure 8.4a), which is a unique behavior compared to that of K, Mn, or Zn-promoted catalysts, where both olefins and paraffins simultaneously increase. Figure 8.4b shows the reaction path of the catalyst, where the C_2–C_4 olefins are adsorbed, by the catalyst particles, and further hydrogenated or oligomerized, leading to light and heavy HCs.

The impact of the use of green hydrogen during the conversion of CO_2 to HCs was addressed by Li et al.,[40] who reported the performance of FeMnK/zeolite-based catalysts coupled with a source of green hydrogen based on a solid oxide electrolytic cell (SOEC). The catalyst was preliminary optimized in terms of metal promotion, support topology, and particle size. From nine zeolite topologies studied, HZSM-5 yielded the best results with an excellent outcome in the terms of production of C_5–C_{11} HCs (70% selectivity), an enhanced CO_2 conversion (27%), and olefin selectivity of 17%. Interestingly, a 10 MR opening pore zeolite favors the formation of C_5–C_{11} HCs, 8 and 12 MR opening pores did not promote the chain growth. As the temperature increased to 400°C, which is the working temperature of the SOEC, the CO_2 conversion increased from 32 to 51%, maintaining the CO selectivity at 15%. Aromatics were also formed on the FEMnK + HZSM-5 catalyst. Figure 8.5 depicts the established mechanism of liquid HC formation from this catalyst. CO_2 is converted into CO in the Fe_3O_4 core-shell structure, followed by hydrogenation to C_2–C_4 olefins at the Fe_5C_2 phase; the olefins are diffused through the 10 MR zeolite pores, and they are oligomerized, cyclized, and dehydrogenated to generate aromatics or oligomerized and isomerized to produce iso-paraffins, which is promoted by the Brönsted acid sites of the zeolite. It is evident here that the pore size is a determinant for the successful production of liquid HCs from CO_2. Furthermore, the production of automotive gasoline is absolutely realizable by processes such as that shown here, adapting the conditions of green hydrogen production.

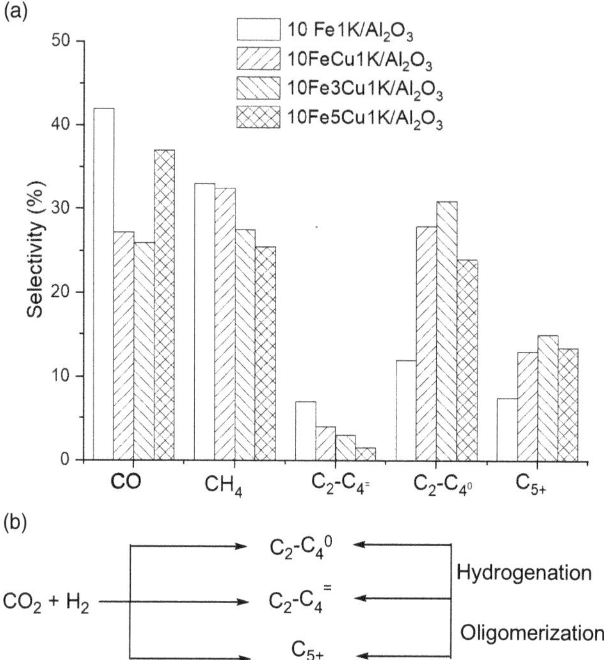

FIGURE 8.4 (a) Selectivity and conversion of hydrocarbons of Cu-promoted K-Fe/Al$_2$O$_3$; (b) reaction path of the FeKCu/Al$_2$O$_3$ catalysts. Adapted with permission from: Liu, J.; Zhang, A.; Jiang, X.; Liu, M.; Sun, Y.; Song, C.; Guo, X. Selective CO$_2$ Hydrogenation to Hydrocarbons on Cu-Promoted Fe-Based Catalysts: Dependence on Cu–Fe Interaction. *ACS Sustain. Chem. Eng.* 2018, *6* (8), 10182–10190. https://doi.org/10.1021/acssuschemeng.8b01491, copyright 2018, American Chemical Society.

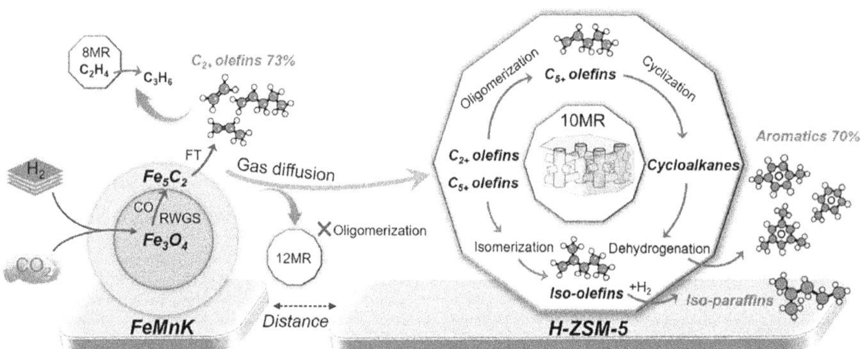

FIGURE 8.5 Mechanism of liquid HCs production from CO$_2$ with FeMnK/H-ZSM-5 catalysts. Reproduced with permission from: Li, Y.; Zeng, L.; Pang, G.; Wei, X.; Wang, M.; Cheng, K.; Kang, J.; Serra, J. M.; Zhang, Q.; Wang, Y. Direct Conversion of Carbon Dioxide into Liquid Fuels and Chemicals by Coupling Green Hydrogen at High Temperature. *Appl. Catal. B Environ.* 2023, *324*, 122299. https://doi.org/10.1016/j.apcatb.2022.122299.

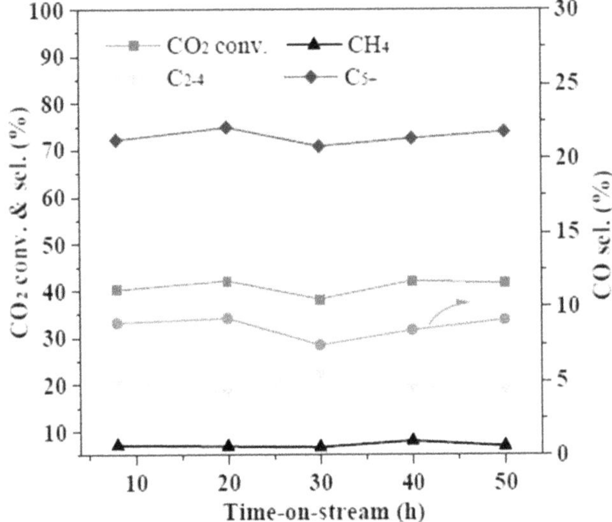

FIGURE 8.6 Conversion and selectivity of the Na-ZnFeCu LDH catalyst. Available via Creative Commons CC licence by Li, Z.; Wa- ang, K.; Xing, Y.; Song, W.; Gao, X.; Ma, Q.; Zhao, T.; Zhang, J. Synthesis of Liquid Hydrocarbon via Direct Hydrogenation of CO_2 over FeCu-Based Bifunctional Catalyst Derived from Layered Double Hydroxides. Molecules 2023, 28 (19), 6920. https://doi.org/10.3390/molecules28196920.

In a completely different mechanism the catalyst reported by Li et al., was reported to act[41]. They reported on layered double hydroxide (LDH)-based FeCu catalysts, which was prepared by the coprecipitation method. These catalyst systems were all promoted by Na and a series of metals such as Al, Mn, Ga, Mg, and Zn. Of them, the Na-MgFeCu LDH exhibited the highest selectivity toward CO and CH_4, but the best performance to produce liquid HCs is shown by the Na-ZnFeCu catalyst, given the better CO_2 adsorption compared to the rest of the synthesized materials. The Na-ZnFeCu material reached a fantastic 72.2% selectivity toward C_{5+} products, at an excellent 40.2% CO_2 conversion. Figure 8.6 shows that a sustained conversion and selectivity are maintained at about the mentioned values for at least 50h time-on-stream. In this case, Mg did not show the same boosting of the catalytic activity by suppressing the methane selectivity because there should be a threshold metal concentration where this effect is possible, and this was overpassed for catalyst preparation.

8.3 GENERAL CONSIDERATIONS ON THE PRODUCTION OF FUELS FROM CO_2

As we have seen before, there are several ways in which CO_2 can be converted into HCs. The CO_2 can be determinant since, among others, it can be obtained as a byproduct of the fermentation process, combined with methane,[42] which in turn can be used as reaction mixture in a process called dry reforming of methane, producing CO and H(subindex)2. This will not be part of this chapter; instead, the reader is referred to some related literature.[43–46] The produced syngas can be further used to generate HCs by the FT process. When only chemical processes are involved, methane is produced initially by the first reduction product of CO_2, CO, by the reactions:

$$CO + 3H_2 \leftrightarrow CH_4 + H_2O \quad \Delta H^0 = -206 \text{ KJ/mol} \tag{8.4}$$

The result of Eq. 8.1 and 8.4 leads to:

$$CO_2 + 4H_2 \leftrightarrow CH_4 + 2H_2O \quad \Delta H^0 = -206 \text{ KJ/mol} \tag{8.5}$$

The importance of both these reactions lies in that they are the reactions associated with the formation of coke and heavy paraffins, being the second ones the target products. In doing so, task-specific catalysts must be designed to produce syngas from methane (dry or steam reforming) and from it, the production of heavy HCs by the FT processes. It should be noted that this indirect method is less convenient in terms of costs and energy consumption. Thus, the direct hydrogenation of CO_2, where the syngas produced is further hydrogenated to C_{5+} HCs, looks more promising but is limited by the poor conversion of CO_2. A second way is to convert CO_2 into methanol, CH_3OH, which can already be used as a high-energy-density fuel or convert it into heavy HCs by the methanol-to-gasoline (MTG) process. As we will see, the production of methanol from CO_2 is also challenging from the point of view of CO_2 conversion. This can be carried out by direct CO_2 hydrogenation (Eq. 8.6) or by hydrogenation of CO (Eq. 8.7).

$$CO_2 + 3H_2 \leftrightarrow CH_3OH + H_2O \quad \Delta H^0 = -49.5 \text{ KJ/mol} \tag{8.6}$$

$$CO + 2H_2 \leftrightarrow CH_3OH \quad \Delta H^0 = -90.6 \text{ KJ/mol} \tag{8.7}$$

8.4 PRODUCTION OF METHANOL FROM CO_2

The methanol produced by the last process (Eq. 8.7) requires more energy than that of direct CO_2 hydrogenation since the raw CO needed must be produced from the RWGS process (Eq. 8.1), which is an endothermic reaction; furthermore, water must be continuously removed to avoid catalyst deactivation and push the reaction toward CO formation, although the whole yield is lower in the direct method due to the high thermodynamic stability of the CO_2, making it much less reactive than CO. As Eqs. 8.6 and 8.7 are both of exothermic nature, and RWGS (Eq 8.1) as well, methanol synthesis from CO_2 is accompanied by a decrease in volume. Therefore, the whole reaction is favored by the application of an increasing pressure and a decreasing temperature, and the equilibrium composition of the feed will determine the maximum potential conversion.

The production of methanol has been long studied and can be carried out by diverse methods. This C_1-compound is much more versatile than CH_4 due to its higher reactivity, mainly due to the presence of one oxygen atom, making possible the preparation of, among others, large HCs, dimethylether (DME), formaldehyde, acetic acid, biodiesel, and chemicals that can be used as fuels, varnishes, paints, solvents, merchandise cleansing products, etc. This versatility, and especially the possibility to industrially convert CO_2 to methanol, made Nobel Prize winner George Olah[47,48] to realize the term "methanol economy" to completely replace fossil fuels and use methanol and DME as sources of energy for transportation, synthesis of HCs, and energy storage. Currently, several methods for the synthesis of methanol from CO_2 have been realized at the laboratory level; to mention, the thermal, hydrogenative, electrochemical, photoelectrochemical, and biochemical conversion methods have been successfully performed. In this chapter, we will focus on the catalytic, thermal, and hydrogenative conversions since they are currently being applied at the commercial scale. For other synthesis methods, the reader is directed to some related literature.[49–52]

As it has been described in Eqs. 8.6 and 8.7, the reactions needed to produce methanol from CO or CO_2 are both of exothermic nature, although they can hardly be carried out without the promotion of a catalyst. It is thus necessary to use moderate temperatures (~150°C) and apply high pressures (5–10 MPa). Direct methanol production from CO_2 generates water as shown in Eq. 8.6. Therefore, suitable dehydration agents are mandatory to avoid catalyst deactivation or reaction rate retardation and to improve the methanol yield. One way to remove water is by catalytic distillation and the other is the by use of selective membranes such as zeolite-based materials.[53,54] This makes the CO_2 direct method of low economical feasibility. One way to boost the methanol yield is the addition of CO into the reaction, which showed an outstanding effect on the amount of methanol per gram of catalyst.[55] The nature of the catalyst turns out to be essential to the process. Currently, the industrial process works through the application of $Cu/ZnO/Al_2O_3$ catalysts, which acts with Cu as the active metal center, Zn as the promotor, and Al_2O_3 as the support[56] Other supports have also been used, such as silica, ceria, titania, etc.[13] As in many industrial processes, the use of heterogeneous catalysts facilitates its reuse therefore, handling the possibility of using different reactor designs is of vital importance from the technical and economical points of view. In this chapter, only heterogeneous catalyst systems will be addressed. First of all, some overviews of the industrial processes of methanol production from CO_2 are presented, which are followed by examples of catalyst technologies.

8.4.1 Industrial Methanol Production from CO_2

The preparation of methanol from syngas has long been known,[36,57] and the typical catalyst systems used are Cu-ZnO at the industrial scale; however, the application of supports such as Al_2O_3 turned out to be beneficial when CO_2 is used as the raw material. Both processes are interrelated in terms of catalyst nature, reactor design, and reactor kinetics. Thus, the design of a CO_2-active catalyst will definitely be able to enhance the two processes' yield and efficiency. The as-synthesized methanol can also be further reacted to generate DME in a dehydration reaction:

$$2CH_3OH \rightarrow CH_3OCH_3 + H_2O \quad \Delta H^0 = -23.4 \text{ JK/mol} \quad (8.8)$$

Both reactions are propagated in a single-step process, thus, the level of dehydration must be well controlled to avoid the generation of undesired products. The production of methanol from syngas can be boosted by the aggregation of 3 wt.% of CO_2 Its production as fuel from CO_2 at an industrial scale is feasible if proper conditions such as catalyst, reactor configuration, dehydrating agent, and separation method are applied.

The first known process to produce methanol from CO_2 can be found back in the 90s when researchers at Lurgi used the Cu-ZnO-based catalyst (67.4 wt.% CuO,

21.4 wt.% ZnO, 11.1 wt.% Al_2O_3) provided by Clariant (then Süd-Chemie). The feed gas contained a high concentration of CO. This innovative process is composed of two units: WGS and methanol units (Figure 8.7). The feed gas ($CO_2/H_2/CO$) is taken to an adiabatic reactor and contacted to the catalyst, whose products are then taken to a water-cooled methanol reactor. This way, the total piping and steel needed are reduced. Some of the side products are DME, methyl-ethyl ketone (EMK), acetone, and ethanol in concentrations of about 2000 ppm. From them, ketones are the most difficult to separate; however, this process generated much less by-product content than those from syngas synthesis due to the lower temperature needed here compared to that process.[14,58]

After the Lurgi process, Joo et al.[55] reported on the CAMERE process in the late 90s. The CAMERE project was carried out by the Korea Electric Power Research Institute (KEPRI) and the Korean Pohang iron and steel company. Essentially, the process consists of a two-stage reaction scheme (Figure 8.8). In the first stage, the gas feed, CO_2, is transformed into CO and water by the RWGS reaction, whose product ($CO/CO_2/H_2$) is taken to a separator. Water is eliminated for the second stage, and the remaining gas products are carried to the methanol reactor. In this stage, some water is produced, which is removed from the effluent by WGS reaction. Both the RWGS and the methanol reactors were serially installed, which reduced the recycle and the purge gas volume at the same time. This way, more CO_2 is available to enter the reaction stream, and the CO_2 conversion is increased. This plant operates to generate about 0.1 Ton of methanol per day with an overall carbon conversion of 89% compared to

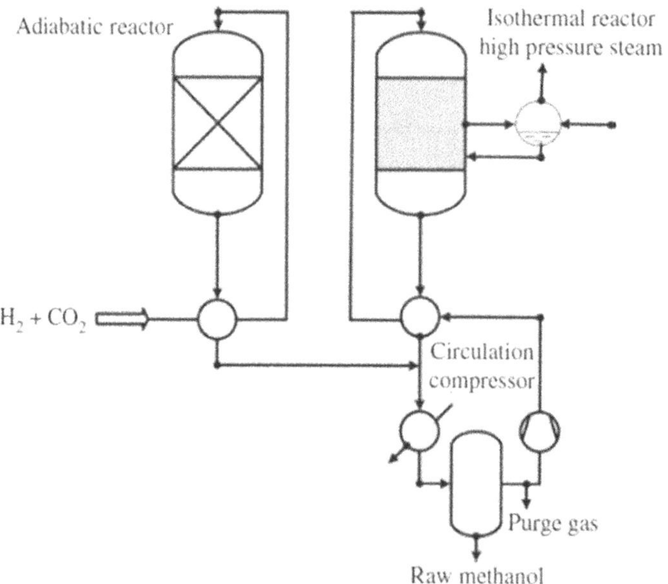

FIGURE 8.7 Lurgi's methanol synthesis process diagram. Reproduced with permission from: Samiee, L.; Gandzha, S. Power to Methanol Technologies via CO_2 Recovery: CO_2 Hydrogenation and Electrocatalytic Routes. *Rev. Chem. Eng.* 2021, *37* (5), 619–641. https://doi.org/10.1515/revce-2019-0012.

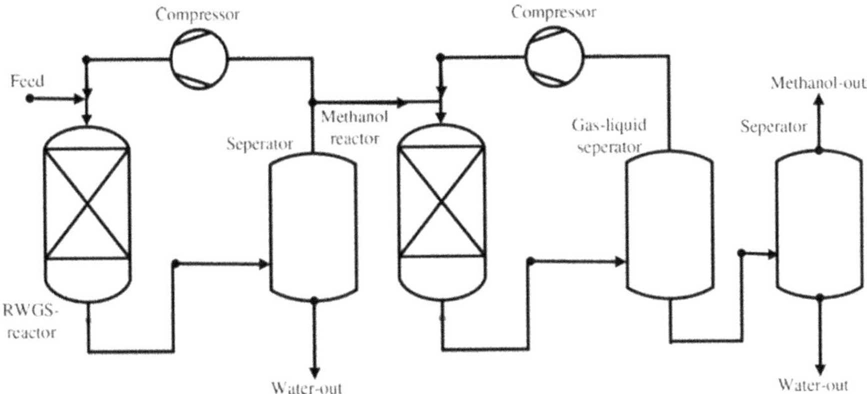

FIGURE 8.8 CAMERE's reaction process scheme. Reproduced with permission from: Samiee, L.; Gandzha, S. Power to Methanol Technologies via CO$_2$ Recovery: CO$_2$ Hydrogenation and Electrocatalytic Routes. *Rev. Chem. Eng.* 2021, *37* (5), 619–641. https://doi.org/10.1515/revce-2019-0012.

the process without RWGS (69%). The catalyst in this process consists of Cu/ZrO$_2$/ZnO/Ga$_2$O$_3$ (5:3:3:1).

One of the processes that became fully operational and gained great success was the Mitsui process (2009) with approximately 0.28 tons/day of methanol production.[59] This process, shown in Figure 8.9, consists of one CO$_2$ and one H$_2$ purification device (1 and 2, respectively). The gases are mixed in a booster compressor (3), followed by contact with the heater (4), where they are heated to the working temperature and finally taken to the reactor (5) for methanol synthesis. The products are cooled with a condenser (6) separated by the devices (7) for liquid and (8) gas products. Part of the gaseous products are recirculated to the reactor (5) by the circulating compressor (9). The process is possible thanks to a catalyst of the composition CuO 45.2 wt.%, ZnO 27.1 wt.%, Al$_2$O$_3$ 4.5 wt.%, ZrO$_2$ 22.6 wt.% and SiO$_2$ 0.6 wt.% that works under a pressure of 5.0 MPa, at 250°C, and 1000 hr^{-1} GHSV at 16.6 L catalyst packing scale.

In Iceland, the Carbon Recycling International (CRI) process was developed[60] using the benefits from the geographical location of the country, where geothermal fields are found. Therefore, the energy that the plant uses is completely clean.[61] The plant, thus, produces green hydrogen for the process (water electrolysis) and uses CO$_2$ emissions from the same geothermal power plants. The process works under the action of a Cu/ZnO/Al$_2$O$_3$, Cu/ZrO$_2$, or a Cu-Zn-Zr-Cr catalyst at 5 MPa and 225°C. Not many details of the process have been published; however, it is known that it works in a single tube–cooled reactor. So, multiple tubes are arranged in a catalyst support plate, which is beneficial to improve heat distribution, prevent hotspots, and increase catalyst life. In 2014, the George Olah plant (GO) was scaled up to 10 Tons/day of methanol production, which corresponds to 2% of the Icelandic gasoline consumption of RM fuel. Furthermore, a part of the product mixture is taken to an MTG reactor, where it is converted into C$_5$–C$_{10}$ HC range fuels of high-octane range.[62]

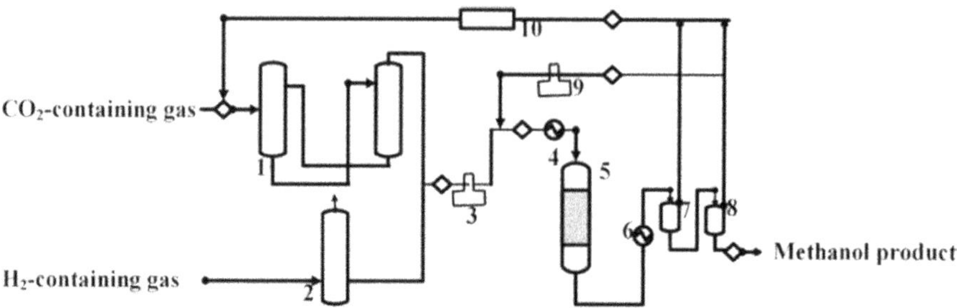

FIGURE 8.9 The Mitsui process scheme. Reproduced with permission from: Samiee, L.; Gandzha, S. Power to Methanol Technologies via CO_2 Recovery: CO_2 Hydrogenation and Electrocatalytic Routes. *Rev. Chem. Eng.* 2021, *37* (5), 619–641. https://doi.org/10.1515/revce-2019-0012.

The MefCO₂ project is a huge endeavor of the European Union to develop an efficient process to capture and convert CO_2 to methanol using H_2 from water electrolysis. This process is aimed to be applied jointly with the production of electricity/heat from biomass and coal power plants, which will bring advantages for exhaust gas transportation (Figure 8.10). The electrolysis energy input will be supplied by surplus energy that can hardly be integrated into the power plants. Then, green hydrogen is intended to be applied. So far, the pilot plant is producing 1 Ton/day of methanol, using around 1.5 Ton of CO_2, tested already with more than 60 research catalysts at an increased activity and selectivity. The plan is to offset surplus CO_2 with a high demand for methanol and contribute to the CCS business case by coupling it with green methanol and reaching the 14% target of renewable fuels in transportation by 2030.

8.4.2 MECHANISM OF METHANOL SYNTHESIS FROM CO_2

As mentioned above, methanol synthesis with the system $Cu/ZnO/Al_2O_3$ is being applied at the industrial scale. This process allows the use of zeolite membrane reactors, which enable the elimination of water. By far, the most studied metal for CO_2 reduction reactions is Cu, mainly due to its high selectivity toward MeOH and whose reaction mechanism has been well studied, such as the work of Higham et al.,[63] who used density functional theory (DFT) (VASP code, version, 5.4.4) to study the effect of CO_2 into Cu(110) and Cu(100) surfaces. Figure 8.11 depicts their findings on the possible reaction pathways for methanol synthesis. Overall, the reaction involves the formation of three C-H bonds, one O-H bond, and the rupture of one of the C-O bonds of the CO_2. Furthermore, the proposed mechanism involves the formation of several intermediate species such as H_2COOH^*, $HCOOH^*$, CH_2OH^*, H_2CO^*, H_2COO^*, and CH_3O^*. The process can start under two scenarios: raw CO_2 being hydrogenated whether to $HCOO^*$ or $COOH^*$ or its dissociation and subsequent hydrogenation to CO^* (activation barrier of 0.138 eV on Cu(100)). This important contribution opens the possibility of designing task-specific catalysts that enable more efficient production of methanol from CO_2.

Some others suggest a mechanism where the RWGS reaction is predominant, such as the proposals made independently by Grabow Liu,[64,65] both defending the

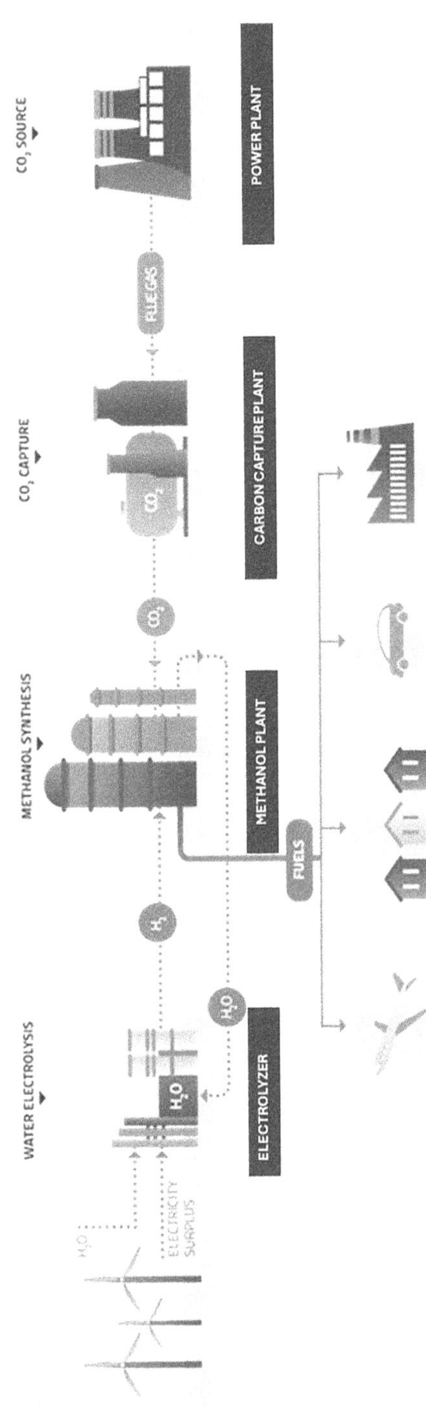

FIGURE 8.10 The MefCO$_2$ concept. Reproduced with permission from: Samiee, L.; Gandzha, S. Power to Methanol Technologies via CO$_2$ Recovery: CO$_2$ Hydrogenation and Electrocatalytic Routes. *Rev. Chem. Eng.* 2021, 37 (5), 619–641. https://doi.org/10.1515/revce-2019-0012.

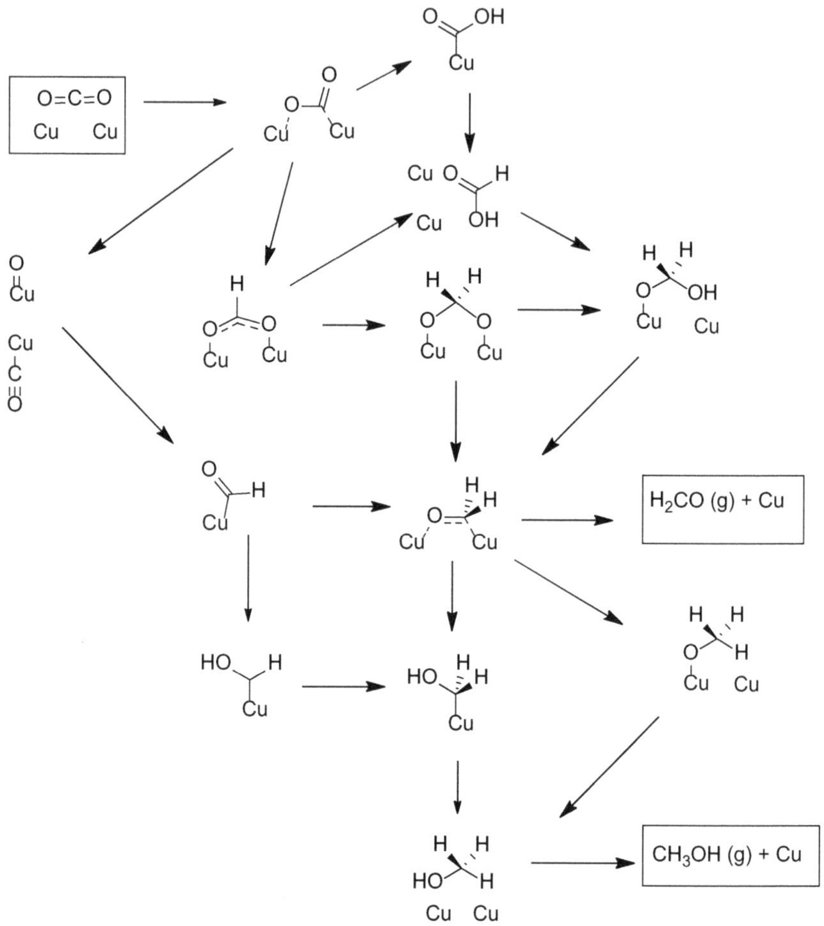

FIGURE 8.11 Reaction pathways for CO_2 hydrogenation over Cu catalysts. Reproduced from: Higham, M. D.; Quesne, M. G.; Catlow, C. R. A. Mechanism of CO_2 Conversion to Methanol over Cu(110) and Cu(100) Surfaces. *Dalton Trans.* 2020, *49* (25), 8478–8497. https://doi.org/10.1039/D0DT00754D. With permission from the Royal Society of Chemistry.

pathway of the RWGS reaction as the early stage of the CO_2 reduction mechanism. This RWGS-based mechanism explains the formation of several intermediates such as formic acid (HCOOH) and hydroxymethyl (CH_3O_2) and permits the production of formaldehyde (CH_2OH), formic acid (HCOOH), and methylformate ($HCOOCH_3$) as side products. Liu suggests that, besides the Cu^0 species participating, Cu^{+1} species are present and take a special role in CO adsorption, making it the main carbon source of the reaction. Thus, the redox reactions taking part are:

$$CO_2 + 2\ Cu^0 \leftrightarrow Cu_2O + CO \tag{8.9}$$

$$H_2 + Cu_2O \leftrightarrow 2Cu^0 + H_2O \tag{8.10}$$

The Cu^0 species induce the dissociation of CO_2 by the oxidation reaction (Eq. 8.9) and the reducing agent, H_2, produces water as a side product, not directly participating as the CO_2 reduction agent. This RWGS approach has been accepted by several authors, some of them including the effect of the supporting material. The reader can refer to the related studies.[66–68]

8.4.3 Progress in Catalyst Systems for the Transformation of CO_2 into Methanol

In the quest for the most active, stable, and selective catalyst for the direct hydrogenation of CO_2 to produce methanol, many synthetic approaches have been realized, with most of them having Cu as the active-phase component of the catalyst system, although the rest of group I transition metals (Au and Ag), Pd, Pt, Ni, Ga, as well as In have also been tried.[69] These catalysts are normally composed of an active phase, promoter, support, and, for industrial purposes, a binder material. It is well known that the interactions between the catalyst components play a key role in the improvement/worsening of the catalyst activity; to name some of them, metal–support interaction, electronic promotion, particle size control, and cocatalyst effect, all of them are essential parameters to consider when designing catalytic materials and can be achieved by a good choice of the synthesis method, adequate support, addition of a metal promoter, and/or an effective cocatalyst. In the following section, some examples of how these parameters apply in the design of catalysts for CO_2 conversion into methanol are presented and how they have been addressed.

There are many examples of the modification of the catalyst composition that enable the tuning of the catalyst properties for the best performance toward the CO_2 reduction reactions. Although Cu can be a very active metal, it can also be poorly active if no support or a nonproper support is used. It has been found, for example, that the use of Cu with no support can yield up to 10^{-8} Kg/m^2 cat/h; however, the use of ZnO/CuO supported on Cu surface generates 3.6 x 10^{-5} kg/m^2 cat/h. Interesting studies of the effect of support have been carried out by the group of Li et al.,[70] by comparing the catalyst activity of three systems, Cu/Al_2O_3, Cu/CeO_2, and Cu/AlCeO, prepared by the coprecipitation method, where the best performance was that of the Cu/AlCeO catalyst, with a 24% CO_2 conversion at 280°C, although the selectivity dramatically decreased to 22% from 85% at 200°C. The finding was that both Al and Ce inhibit Cu crystallite growth and CeO_2 hindered the agglomeration of the Cu particles, thus controlling the size of the Cu particle at a minimum, which was beneficial for the adsorption of CO_2. CeO_2 also increased surface basicity and the atomic ratio of Cu^+ ions. Furthermore, the activation energy of the Cu/AlCeO catalyst turned out to be much lower than the other two composites, demonstrating the synergistic effect of both support components. Chang et al.[71] made a very important contribution to the study of the interaction between the support components. They reported on mixed oxides of CuCe, CuTi, and CuCeTi synthesized by the sol-gel and coprecipitation methods. The results of the catalytic activity demonstrated that the ternary oxide CuCeTi was more active than the binary systems (Figure 8.12). The reactions were performed at 3 MPa, 235°C, at a gas hourly space velocity (GHSV) of 2000 mL/g cat/h. The analysis revealed that the ternary oxide system ($CuCeTiO_x$) had the largest

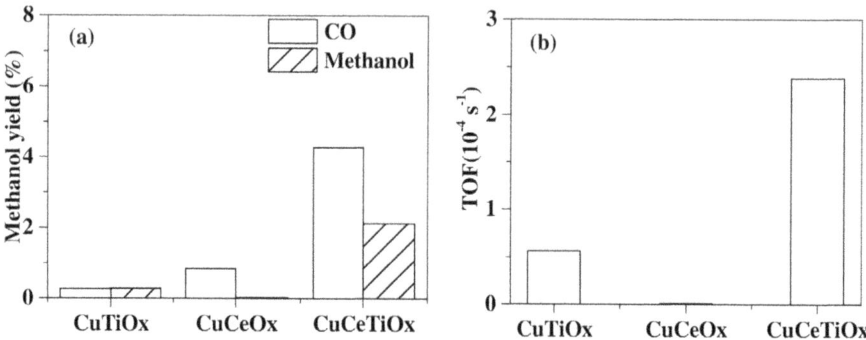

FIGURE 8.12 Methanol yield (a) and TOF (a) for the catalysts $CuTiO_x$, $CuCeO_x$, and $CuCeTiO_x$ catalysts. Reproduced with permission from: Chang, K.; Wang, T.; Chen, J. G. Hydrogenation of CO_2 to Methanol over CuCeTiOx Catalysts. *Appl. Catal. B Environ.* 2017, *206*, 704–711. https://doi.org/10.1016/j.apcatb.2017.01.076.

surface area and the higher number of oxygen vacancies, which facilitated CO_2 intake by the Cu surface. When comparing the synthesis method, the sol-gel synthesis method resulted in a higher catalyst selectivity but lower CO_2 conversion. Here, the Ce-to-Ti ratio was relevant, since they suppress each other from sintering, resulting in the 1:1 ratio being the optimal one.

When referring to the metal–support interaction, this has also critical repercussions on the catalytic activity. As an example, Wang et al.[72] studied two catalyst systems Cu/CeO_2 and Cu/ZrO_2 prepared by the oxalate precipitation method. The CO_2 reduction reactions were carried out at 3 MPa, 200°C, using a H_2/CO_2 of 3:1. The supports clearly enhanced the catalytic activity of Cu, but the effect of Ce on the conversion and selectivity was higher than that of Zr. The Raman spectra revealed that when using Cu/CeO_2, more oxygen vacancies are formed. Furthermore, these vacancies are converted into carbonate species, and, as a result, higher amounts of $HCOO^-$ and CH_3O^- are produced. The key intermediates in the formation of methanol, thus improving the selectivity of the Cu/CeO_2 catalyst.

One good analysis of the catalyst preparation method was conducted by Dasireddy et al.[73] Their group used four different methods (coprecipitation, ultrasound-assisted, sol-gel, and solid-state) to prepare $Cu/ZnO/Al_2O_3$ catalysts. Each of the preparation methods had an important influence on the Cu particle size, Cu^0/Cu^+ ratio, and, of course, the distribution of the Cu particles on the support's surface. From all the studied methods (Figure 8.13), the ultrasound-assisted synthesis yielded the best selectivity at lower temperatures due to the higher amount of basic active sites generated and the higher dispersion of Cu particles, modifying the CuO–ZnO interaction and, thus, its reactivity. It was also found that the presence of high-concentration Cu^+ species favored selectivity and that the catalyst activity has an intimate relationship with the surface area of Cu, which is almost linear.

The effect of metal promoters on Cu catalysts has also been widely studied. Ladera et al.[74] reported on the effect of Ga on the $Cu/ZnO/ZrO_2$ catalyst (CZZ). The addition of Ga_2O_3 was found to increase the distribution of Cu due to the structural promotion of Ga, enhancing the sintering resistance of CuO during the thermal treatment. The

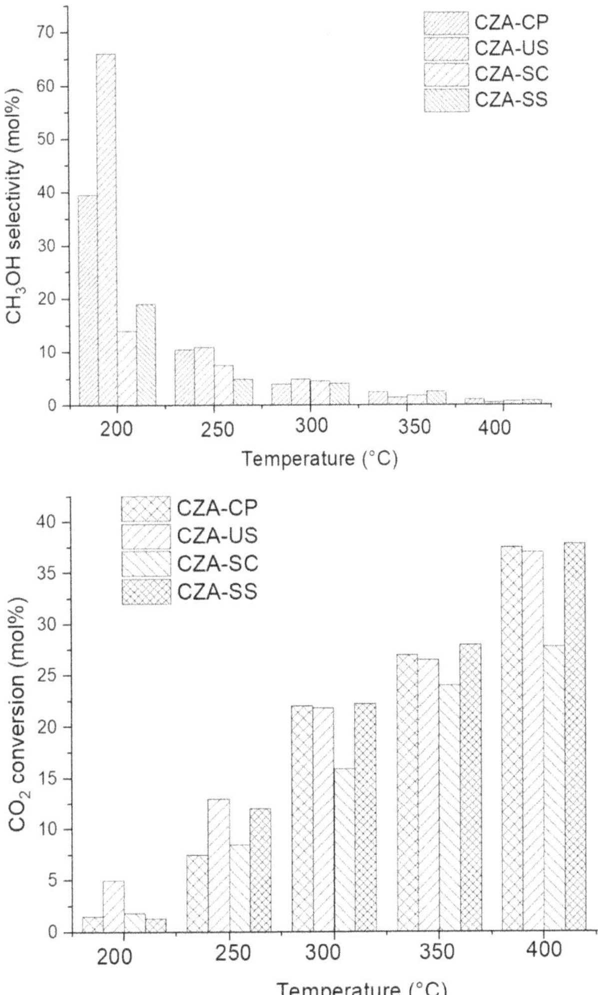

FIGURE 8.13 Comparison of the conversion (A) and selectivity (B) of $Cu/ZnO/Al_2O_3$ (CZA) catalysts in terms of the synthesis method (CP: coprecipitation, US: ultrasound-assisted precipitation, SC: sol-gel combustion, SS: solid state). Reproduced with permission from: Dasireddy, V. D. B. C.; Likozar, B. The Role of Copper Oxidation State in Cu/ZnO/Al2O3 Catalysts in CO_2 Hydrogenation and Methanol Productivity. *Renew. Energy* 2019, *140*, 452–460. https://doi.org/10.1016/j.renene.2019.03.073.

catalysts were prepared by the coprecipitation method and the catalysis was measured at the temperature range of 190–300°C, at pressures 3–7 MPa, a H_2/CO_2 molar ratio of 3:1, and GHSV of 15,000 mL/g cat/h. The Ga_2O_3 effect on the surface Cu concentration was reflected in the higher activity toward CO_2 reduction. In general, the CO_2 conversion rate increases with the temperature, conversely to the methanol selectivity.

As mentioned before, there are quite a large number of catalyst systems that have proven activity toward methanol synthesis. Table 8.3 gathers some examples of

TABLE 8.3

Reaction Conditions and Outcome of Some of the Catalysts Used for Direct Conversion of CO_2 to Methanol[13,69,75]

Catalyst	CO_2/H_2	P (MPa)	T(°C)	WHSV/GHSV(mL $g_{cat}^{-1}h^{-1}$)	CO_2 (%conv)	MeOH (%sel.)	STY ($g_{MeOH}g_{cat}^{-1}h^{-1}$)	Ref.
Cu/ZrO	1:3	8	250	3600	15	86	—	76
Cu/ZnO/Al$_2$O$_3$	1:3	5	270	4000	23.7	43.7	0.15[b]	77
Cu/ZnO/Al$_2$O$_3$/ZrO$_2$	1:3	5	190	4000	10.7	81.8	0.087	78
Cu/ZnO/Al$_2$O$_3$/Y$_2$O$_3$	1:3	9	230	10000	29.9	89.7	0.39	79
Cu/ZnO/Al$_2$O$_3$/Ga$_2$O$_3$	1:2.8	4.5	240	18000	27	50	—	80
CuZnO/SiO$_2$	1:3	3	220	2000 mL $g_{cat}^{-1}h^{-1}$	14.1	57.2	0.0554	81
CuO-ZnO-ZrO$_2$ (1.0 wt% SiO$_2$)	1:3	2	240	3900 mL $g_{cat}^{-1}h^{-1}$	5	70	0.25	82
CuZnAlZr (6:3:0.5:0.5)	1:3	5	270	4600 h^{-1}	24.5	57.6	0.21	83
Rh/TiO$_2$	1:4	0.1	350	8500 mL $g_{cat}^{-1}h^{-1}$	66	100	—	84
Cu/ZnCr	1:3	2	300	6000 mL $g_{cat}^{-1}h^{-1}$	25.1	31.1	0.15	85
Cu/Zn/Al (66/30/11)	1:3	2.8	220	1525 mL $g_{cat}^{-1}h^{-1}$	20.3	63.2	—	86
AuCuO/CeO$_2$	1:3	3	240	—	6.7	29.6	—	87
CuO-ZnO/Al$_2$O$_3$	1:6	1	227	684 mL $g_{cat}^{-1}h^{-1}$	13.7	74	—	88
Cu/AlCeO	1:3	3	260	14,400	17	45	0.381	70
CuZnZr/CuBr$_2$	1:3	5	250	3000	10.7	97.1	0.1	89
CuZnAlCe	1:3	3	250	12,000	14.2	37.8	0.213	90
Cu-Zn-Al-K	1:4	3	240	2,400	14	96	0.461	91
Cu/Zn/Al/Zr (52.5:24.9:17.1:5.5:3.33)	1:3	5	230	8500	19.3	58.5	0.33	92
Cu/Zn/Al/Zr/Fe (52.8:24.6:17.1:5.5:3.33)	1:3	5	230	8500	18	68.4	0.32	92
Cu-Zn/Al foam	1:3	3	250	20000	13.6	64.5	7.81	93
CuZnCr	1:3	2	300	6000	25.1	31.1	0.15	85

Catalyst								
In$_2$O$_3$/Co$_3$O$_4$	1:3	5	285	15,600	17.3	75	0.65	94
Ir/In$_2$O$_3$	1:4	5	300	21,000	17.7	70	0.765	95
Au/In$_2$O$_3$-ZrO$_2$	1:4	5	300	21,000	14.8	70.1	0.59	96
Rh/In$_2$O$_3$	1:4	5	300	21,000	17.1	56.1	0.5448	97
ReO$_x$/TiO$_2$	1:4	10	200	4	18	98	7.9113	98
Pd/Zn/ZnO/ZnFe$_2$O$_4$	1:3	5	290	21,600	13.94	55.02	0.953	99
MoS$_2$	1:3	5	180	3000	12.5	94.3	0.13	100
Pd/ZnO	1:3	2	250	3600	11.1	59	—	101
Pd/ZnO/TiO$_2$	1:3	2	250	3600	10.1	40	—	102
Pd/plate Ga$_2$O$_3$	1:3	5	250	60000	17.3	51.6	—	103

catalysts that have been employed for this purpose, choosing those that have shown more than 10% of CO_2 conversion or more than 50% of methanol selectivity. The literature on the topic has increased in recent years; due to the growing interest in GHG emission reduction. The term "methanol economy" is a potential reality since the carbon recycling industry is making this "dream" to come true.

8.5 GENERAL OUTLOOK AND PERSPECTIVES

The use of fossil fuels as a source of automotive gasoline is going to be just part of human history. There is enormous pressure on the oil and gas industry to reduce their CO_2 emissions, which has led many companies to develop ambitious research efforts to reduce carbon emissions to the atmosphere. Their great advantage is the fact that the combustion engines for automobiles will hardly be eliminated in the future decades, especially in underdeveloped countries, where investment in electric fuels is not an option because of the high costs of electric cars and photocell technologies to obtain light. Thus, it is expected that the generation of fuels will continue to be a need in the future. This way, GHGs such as CO_2 will play important roles in the production of fuels, where new developments in the production of HC fuels such as paraffins, olefins, and aromatics will surely come to light. The technology to produce methanol from CO_2, whether to be used as a fuel or as a source of chemicals is much more developed and, as we have seen, is already commercial in some countries. This chapter gives just a small outlook of the research that has been done, progress achieved, and the existing commercial technologies in this respect. The reader is suggested to keep looking for information in the field.

REFERENCES

(1) *UN Climate Change Conference - United Arab Emirates|UNFCCC.* https://unfccc.int/cop28 (accessed January 15, 2024).

(2) Jain, P. C. Greenhouse Effect and Climate Change: Scientific Basis and Overview. *Renew. Energy* 1993, *3* (4), 403–420. https://doi.org/10.1016/0960-1481(93)90108-S

(3) Dubey, A.; Arora, A. Advancements in Carbon Capture Technologies: A Review. *J. Clean. Prod.* 2022, *373*, 133932. https://doi.org/10.1016/j.jclepro.2022.133932

(4) Yusuf, M.; Ibrahim, H. A Comprehensive Review on Recent Trends in Carbon Capture, Utilization, and Storage Techniques. *J. Environ. Chem. Eng.* 2023, *11* (6), 111393. https://doi.org/10.1016/j.jece.2023.111393

(5) Liu, E.; Lu, X.; Wang, D. A Systematic Review of Carbon Capture, Utilization and Storage: Status, Progress and Challenges. *Energies* 2023, *16* (6), 2865. https://doi.org/10.3390/en16062865

(6) Bahman, N.; Al-Khalifa, M.; Al Baharna, S.; Abdulmohsen, Z.; Khan, E. Review of Carbon Capture and Storage Technologies in Selected Industries: Potentials and Challenges. *Rev. Environ. Sci. Biotechnol.* 2023, *22* (2), 451–470. https://doi.org/10.1007/s11157-023-09649-0

(7) Sakakura, T.; Choi, J.-C.; Yasuda, H. Transformation of Carbon Dioxide. *Chem. Rev.* 2007, *107* (6), 2365–2387. https://doi.org/10.1021/cr068357u

(8) Yusuf, N.; Almomani, F.; Qiblawey, H. Catalytic CO_2 Conversion to C1 Value-Added Products: Review on Latest Catalytic and Process Developments. *Fuel* 2023, *345*, 128178. https://doi.org/10.1016/j.fuel.2023.128178

(9) Song, Q.-W.; Ma, R.; Liu, P.; Zhang, K.; He, L.-N. Recent Progress in CO$_2$ Conversion into Organic Chemicals by Molecular Catalysis. *Green Chem.* 2023, *25* (17), 6538–6560. https://doi.org/10.1039/D3GC01892J

(10) Saravanan, A.; Senthil Kumar, P.; Vo, D.-V. N.; Jeevanantham, S.; Bhuvaneswari, V.; Anantha Narayanan, V.; Yaashikaa, P. R.; Swetha, S.; Reshma, B. A Comprehensive Review on Different Approaches for CO$_2$ Utilization and Conversion Pathways. *Chem. Eng. Sci.* 2021, *236*, 116515. https://doi.org/10.1016/j.ces.2021.116515

(11) Artz, J.; Müller, T. E.; Thenert, K.; Kleinekorte, J.; Meys, R.; Sternberg, A.; Bardow, A.; Leitner, W. Sustainable Conversion of Carbon Dioxide: An Integrated Review of Catalysis and Life Cycle Assessment. *Chem. Rev.* 2018, *118* (2), 434–504. https://doi.org/10.1021/acs.chemrev.7b00435

(12) Okoye-Chine, C. G.; Otun, K.; Shiba, N.; Rashama, C.; Ugwu, S. N.; Onyeaka, H.; Okeke, C. T. Conversion of Carbon Dioxide into Fuels—A Review. *J. CO$_2$ Util.* 2022, *62*, 102099. https://doi.org/10.1016/j.jcou.2022.102099

(13) Atsbha, T. A.; Yoon, T.; Seongho, P.; Lee, C.-J. A Review on the Catalytic Conversion of CO$_2$ Using H$_2$ for Synthesis of CO, Methanol, and Hydrocarbons. *J. CO$_2$ Util.* 2021, *44*, 101413. https://doi.org/10.1016/j.jcou.2020.101413

(14) Samiee, L.; Gandzha, S. Power to Methanol Technologies via CO$_2$ Recovery: CO$_2$ Hydrogenation and Electrocatalytic Routes. *Rev. Chem. Eng.* 2021, *37* (5), 619–641. https://doi.org/10.1515/revce-2019-0012

(15) Stöcker, M. Methanol-to-Hydrocarbons: Catalytic Materials and Their behavior. *Microporous Mesoporous Mater.* 1999, *29* (1), 3–48. https://doi.org/10.1016/S1387-1811(98)00319-9

(16) Chang, C. D. Hydrocarbons from Methanol. *Catal. Rev. Sci. Eng.* 1983. https://doi.org/10.1080/01614948308078874

(17) Chakraborty, J. P.; Singh, S.; Maity, S. K. Chapter 6 - Advances in the Conversion of Methanol to Gasoline. In *Hydrocarbon Biorefinery*; Maity, S. K., Gayen, K., Bhowmick, T. K., Eds.; Elsevier, 2022; pp 177–200. https://doi.org/10.1016/B978-0-12-823306-1.00008-X

(18) Santos, M. F.; Bresciani, A. E.; Ferreira, N. L.; Bassani, G. S.; Alves, R. M. B. Carbon Dioxide Conversion via Reverse Water-Gas Shift Reaction: Reactor Design. *J. Environ. Manage.* 2023, *345*, 118822. https://doi.org/10.1016/j.jenvman.2023.118822

(19) González-Castaño, M.; Dorneanu, B.; Arellano-García, H. The Reverse Water Gas Shift Reaction: A Process Systems Engineering Perspective. *React. Chem. Eng.* 2021, *6* (6), 954–976. https://doi.org/10.1039/D0RE00478B

(20) Jahangiri, H.; Bennett, J.; Mahjoubi, P.; Wilson, K.; Gu, S. A Review of Advanced Catalyst Development for Fischer–Tropsch Synthesis of Hydrocarbons from Biomass Derived Syn-Gas. *Catal. Sci. Technol.* 2014, *4* (8), 2210–2229. https://doi.org/10.1039/C4CY00327F

(21) Gao, Y.; Shao, L.; Yang, S.; Hu, J.; Zhao, S.; Dang, J.; Wang, W.; Yan, X.; Yang, P. Recent Advances in Iron-Based Catalysts for Fischer–Tropsch to Olefins Reaction. *Catal. Commun.* 2023, *181*, 106720. https://doi.org/10.1016/j.catcom.2023.106720

(22) Suo, Y.; Yao, Y.; Zhang, Y.; Xing, S.; Yuan, Z.-Y. Recent Advances in Cobalt-Based Fischer-Tropsch Synthesis Catalysts. *J. Ind. Eng. Chem.* 2022, *115*, 92–119. https://doi.org/10.1016/j.jiec.2022.08.026

(23) Chan, Y. H.; Syed Abdul Rahman, S. N. F.; Lahuri, H. M.; Khalid, A. Recent Progress on CO-Rich Syngas Production via CO$_2$ Gasification of Various Wastes: A Critical Review on Efficiency, Challenges and Outlook. *Environ. Pollut.* 2021, *278*, 116843. https://doi.org/10.1016/j.envpol.2021.116843

(24) Aresta, M. Carbon Dioxide Reduction to C1 or Cn Molecules. In *Carbon Dioxide Recovery and Utilization*; Aresta, M., Ed.; Springer Netherlands: Dordrecht, 2003; pp 293–312. https://doi.org/10.1007/978-94-017-0245-4_12

(25) Dorner, R. W.; Hardy, D. R.; Williams, F. W.; Davis, B. H.; Willauer, H. D. Influence of Gas Feed Composition and Pressure on the Catalytic Conversion of CO_2 to Hydrocarbons Using a Traditional Cobalt-Based Fischer–Tropsch Catalyst. *Energy Fuels* 2009, *23* (8), 4190–4195. https://doi.org/10.1021/ef900275m

(26) Iloy, R. A.; Jalama, K. Effect of Operating Temperature, Pressure and Potassium Loading on the Performance of Silica-Supported Cobalt Catalyst in CO_2 Hydrogenation to Hydrocarbon Fuel. *Catalysts* 2019, *9* (10), 807. https://doi.org/10.3390/catal9100807

(27) Shi, Z.; Yang, H.; Gao, P.; Li, X.; Zhong, L.; Wang, H.; Liu, H.; Wei, W.; Sun, Y. Direct Conversion of CO_2 to Long-Chain Hydrocarbon Fuels over K–Promoted CoCu/TiO$_2$ Catalysts. *Catal. Today* 2018, *311*, 65–73. https://doi.org/10.1016/j.cattod.2017.09.053

(28) He, Z.; Cui, M.; Qian, Q.; Zhang, J.; Liu, H.; Han, B. Synthesis of Liquid Fuel via Direct Hydrogenation of CO_2. *Proc. Natl. Acad. Sci.* 2019, *116* (26), 12654–12659. https://doi.org/10.1073/pnas.1821231116

(29) Dorner, R. W.; Hardy, D. R.; Williams, F. W.; Willauer, H. D. K and Mn Doped Iron-Based CO_2 Hydrogenation Catalysts: Detection of KAlH4 as Part of the Catalyst's Active Phase. *Appl. Catal. Gen.* 2010, *373* (1), 112–121. https://doi.org/10.1016/j.apcata.2009.11.005

(30) Dorner, R. W.; Hardy, D. R.; Williams, F. W.; Willauer, H. D. Catalytic CO_2 Hydrogenation to Feedstock Chemicals for Jet Fuel Synthesis Using Multi-Walled Carbon Nanotubes as Support. In *Advances in CO$_2$ Conversion and Utilization; ACS Symposium Series*; American Chemical Society, 2010; Vol. 1056, pp 125–139. https://doi.org/10.1021/bk-2010-1056.ch008

(31) Albrecht, M.; Rodemerck, U.; Schneider, M.; Bröring, M.; Baabe, D.; Kondratenko, E. V. Unexpectedly Efficient CO_2 Hydrogenation to Higher Hydrocarbons over Non-Doped Fe_2O_3. *Appl. Catal. B Environ.* 2017, *204*, 119–126. https://doi.org/10.1016/j.apcatb.2016.11.017

(32) Amoyal, M.; Vidruk-Nehemya, R.; Landau, M. V.; Herskowitz, M. Effect of Potassium on the Active Phases of Fe Catalysts for Carbon Dioxide Conversion to Liquid Fuels through Hydrogenation. *J. Catal.* 2017, *348*, 29–39. https://doi.org/10.1016/j.jcat.2017.01.020

(33) Meiri, N.; Dinburg, Y.; Amoyal, M.; Koukouliev, V.; Nehemya, R. V.; Landau, M. V.; Herskowitz, M. Novel Process and Catalytic Materials for Converting CO_2 and H_2 Containing Mixtures to Liquid Fuels and Chemicals. *Faraday Discuss.* 2015, *183* (0), 197–215. https://doi.org/10.1039/C5FD00039D

(34) Choi, Y. H.; Jang, Y. J.; Park, H.; Kim, W. Y.; Lee, Y. H.; Choi, S. H.; Lee, J. S. Carbon Dioxide Fischer-Tropsch Synthesis: A New Path to Carbon-Neutral Fuels. *Appl. Catal. B Environ.* 2017, *202*, 605–610. https://doi.org/10.1016/j.apcatb.2016.09.072

(35) Wei, J.; Ge, Q.; Yao, R.; Wen, Z.; Fang, C.; Guo, L.; Xu, H.; Sun, J. Directly Converting CO_2 into a Gasoline Fuel. *Nat. Commun.* 2017, *8* (1), 15174. https://doi.org/10.1038/ncomms15174

(36) Biswal, T.; Shadangi, K. P.; Sarangi, P. K.; Srivastava, R. K. Conversion of Carbon Dioxide to Methanol: A Comprehensive Review. *Chemosphere* 2022, *298*, 134299. https://doi.org/10.1016/j.chemosphere.2022.134299

(37) Zhang, X.; Zhang, G.; Song, C.; Guo, X. Catalytic Conversion of Carbon Dioxide to Methanol: Current Status and Future Perspective. *Front. Energy Res.* 2021, 8.

(38) Azhari, N. J.; Erika, D.; Mardiana, S.; Ilmi, T.; Gunawan, M. L.; Makertihartha, I. G. B. N.; Kadja, G. T. M. Methanol Synthesis from CO_2: A Mechanistic Overview. *Results Eng.* 2022, *16*, 100711. https://doi.org/10.1016/j.rineng.2022.100711

(39) Liu, J.; Zhang, A.; Jiang, X.; Liu, M.; Sun, Y.; Song, C.; Guo, X. Selective CO_2 Hydrogenation to Hydrocarbons on Cu-Promoted Fe-Based Catalysts: Dependence on Cu–Fe Interaction. *ACS Sustain. Chem. Eng.* 2018, *6* (8), 10182–10190. https://doi.org/10.1021/acssuschemeng.8b01491

(40) Li, Y.; Zeng, L.; Pang, G.; Wei, X.; Wang, M.; Cheng, K.; Kang, J.; Serra, J. M.; Zhang, Q.; Wang, Y. Direct Conversion of Carbon Dioxide into Liquid Fuels and Chemicals by Coupling Green Hydrogen at High Temperature. *Appl. Catal. B Environ.* 2023, *324*, 122299. https://doi.org/10.1016/j.apcatb.2022.122299

(41) Li, Z.; Wang, K.; Xing, Y.; Song, W.; Gao, X.; Ma, Q.; Zhao, T.; Zhang, J. Synthesis of Liquid Hydrocarbon via Direct Hydrogenation of CO$_2$ over FeCu-Based Bifunctional Catalyst Derived from Layered Double Hydroxides. *Molecules* 2023, *28* (19), 6920. https://doi.org/10.3390/molecules28196920

(42) Goli, A.; Shamiri, A.; Talaiekhozani, A.; Eshtiaghi, N.; Aghamohammadi, N.; Aroua, M. K. An Overview of Biological Processes and Their Potential for CO$_2$ Capture. *J. Environ. Manage.* 2016, *183*, 41–58. https://doi.org/10.1016/j.jenvman.2016.08.054

(43) Ranjekar, A. M.; Yadav, G. D. Dry Reforming of Methane for Syngas Production: A Review and Assessment of Catalyst Development and Efficacy. *J. Indian Chem. Soc.* 2021, *98* (1), 100002. https://doi.org/10.1016/j.jics.2021.100002

(44) Pakhare, D.; Spivey, J. A Review of Dry (CO$_2$) Reforming of Methane over Noble Metal Catalysts. *Chem. Soc. Rev.* 2014, *43* (22), 7813–7837. https://doi.org/10.1039/C3CS60395D

(45) Arora, S.; Prasad, R. An Overview on Dry Reforming of Methane: Strategies to Reduce Carbonaceous Deactivation of Catalysts. *RSC Adv.* 2016, *6* (110), 108668–108688. https://doi.org/10.1039/C6RA20450C

(46) Wittich, K.; Krämer, M.; Bottke, N.; Schunk, S. A. Catalytic Dry Reforming of Methane: Insights from Model Systems. *ChemCatChem* 2020, *12* (8), 2130–2147. https://doi.org/10.1002/cctc.201902142

(47) Olah, G. A. Beyond Oil and Gas: The Methanol Economy. *Angew. Chem. Int. Ed.* 2005, *44* (18), 2636–2639. https://doi.org/10.1002/anie.200462121

(48) Front Matter. In *Beyond Oil and Gas: The Methanol Economy*; John Wiley & Sons, Ltd, 2009; p I–XVI. https://doi.org/10.1002/9783527627806.fmatter

(49) Simon Araya, S.; Liso, V.; Cui, X.; Li, N.; Zhu, J.; Sahlin, S. L.; Jensen, S. H.; Nielsen, M. P.; Kær, S. K. A Review of The Methanol Economy: The Fuel Cell Route. *Energies* 2020, *13* (3), 596. https://doi.org/10.3390/en13030596

(50) Ebrahimzadeh Sarvestani, M.; Norouzi, O.; Di Maria, F.; Dutta, A. From Catalyst Development to Reactor Design: A Comprehensive Review of Methanol Synthesis Techniques. *Energy Convers. Manag.* 2024, *302*, 118070. https://doi.org/10.1016/j.enconman.2024.118070

(51) Lange, J.-P. Methanol Synthesis: A Short Review of Technology Improvements. *Catal. Today* 2001, *64* (1), 3–8. https://doi.org/10.1016/S0920-5861(00)00503-4

(52) Deka, T. J.; Osman, A. I.; Baruah, D. C.; Rooney, D. W. Methanol Fuel Production, Utilization, and Techno-Economy: A Review. *Environ. Chem. Lett.* 2022, *20* (6), 3525–3554. https://doi.org/10.1007/s10311-022-01485-y

(53) Kamkeng, A. D. N.; Wang, M. Technical Analysis of the Modified Fischer-Tropsch Synthesis Process for Direct CO$_2$ Conversion into Gasoline Fuel: Performance Improvement via Ex-Situ Water Removal. *Chem. Eng. J.* 2023, *462*, 142048. https://doi.org/10.1016/j.cej.2023.142048

(54) Desgagnés, A.; Iliuta, M. C. Intensification of CO$_2$ Hydrogenation by In-Situ Water Removal Using Hybrid Catalyst-Adsorbent Materials: Effect of Preparation Method and Operating Conditions on the RWGS Reaction as a Case Study. *Chem. Eng. J.* 2023, *454*, 140214. https://doi.org/10.1016/j.cej.2022.140214

(55) Joo, O.-S.; Jung, K.-D.; Moon, I.; Rozovskii, A. Y.; Lin, G. I.; Han, S.-H.; Uhm, S.-J. Carbon Dioxide Hydrogenation To Form Methanol via a Reverse-Water-Gas-Shift Reaction (the CAMERE Process). *Ind. Eng. Chem. Res.* 1999, *38* (5), 1808–1812. https://doi.org/10.1021/ie9806848

(56) Liang, B.; Ma, J.; Su, X.; Yang, C.; Duan, H.; Zhou, H.; Deng, S.; Li, L.; Huang, Y. Investigation on Deactivation of $Cu/ZnO/Al_2O_3$ Catalyst for CO_2 Hydrogenation to Methanol. *Ind. Eng. Chem. Res.* 2019, *58* (21), 9030–9037. https://doi.org/10.1021/acs.iecr.9b01546

(57) Liu, G.; Hagelin-Weaver, H.; Welt, B. A Concise Review of Catalytic Synthesis of Methanol from Synthesis Gas. *Waste* 2023, *1* (1), 228–248. https://doi.org/10.3390/waste 1010015

(58) Wernicke, H.-J.; Plass, L.; Schmidt, F. Methanol Generation. In *Methanol: The Basic Chemical and Energy Feedstock of the Future*; Springer, Berlin, Heidelberg, 2014; pp 51–301. https://doi.org/10.1007/978-3-642-39709-7_4

(59) Matsushita, T.; Haganuma, T.; Fujita, D. Process for Producing Methanol. US20130 237618A1, September 12, 2013. https://patents.google.com/patent/US20130237618A1/ en (accessed 2024-01-22).

(60) *CRI - Carbon Recycling International*. CRI - Carbon Recycling International. https:// www.carbonrecycling.is (accessed 2024-01-23).

(61) *Production of renewable methanol from captured emissions and renewable energy sources, for its utilisation for clean fuel production and green consumer goods | CIRCLENERGY Project | Fact Sheet | H2020*. CORDIS | European Commission. https://cordis.europa.eu/ project/id/791632 (accessed January 01, 2024).

(62) Shulenberger, A. M.; Jonsson, F. R.; Ingolfsson, O.; Tran, K.-C. Process for Producing Liquid Fuel from Carbon Dioxide and Water. US8198338B2, June 12, 2012. https:// patents.google.com/patent/US8198338B2/en (accessed 2024-01-23).

(63) Higham, M. D.; Quesne, M. G.; Catlow, C. R. A. Mechanism of CO_2 Conversion to Methanol over Cu(110) and Cu(100) Surfaces. *Dalton Trans.* 2020, *49* (25), 8478–8497. https://doi.org/10.1039/D0DT00754D

(64) Grabow, L. C.; Mavrikakis, M. Mechanism of Methanol Synthesis on Cu through CO_2 and CO Hydrogenation. *ACS Catal.* 2011, *1* (4), 365–384. https://doi.org/10.1021/cs200055d

(65) Liu, Y.-M.; Liu, J.-T.; Liu, S.-Z.; Li, J.; Gao, Z.-H.; Zuo, Z.-J.; Huang, W. Reaction Mechanisms of Methanol Synthesis from CO/CO_2 Hydrogenation on Cu_2O(111): Comparison with Cu(111). *J. CO_2 Util.* 2017, *20*, 59–65. https://doi.org/10.1016/j.jcou.2017.05.005

(66) Studt, F.; Behrens, M.; Kunkes, E. L.; Thomas, N.; Zander, S.; Tarasov, A.; Schumann, J.; Frei, E.; Varley, J. B.; Abild-Pedersen, F.; Nørskov, J. K.; Schlögl, R. The Mechanism of CO and CO_2 Hydrogenation to Methanol over Cu-Based Catalysts. *ChemCatChem* 2015, *7* (7), 1105–1111. https://doi.org/10.1002/cctc.201500123

(67) Zhou, H.; Jin, H.; Li, Y.; Li, Y.; Huang, S.; Lin, W.; Chen, W.; Zhang, Y. Mechanism of Methanol Synthesis from CO_2 Hydrogenation over $Cu/\gamma\text{-}Al_2O_3$ Interface: Influences of Surface Hydroxylation. *Catalysts* 2023, *13* (9), 1244. https://doi.org/10.3390/catal 13091244

(68) Zhu, J.; Su, Y.; Chai, J.; Muravev, V.; Kosinov, N.; Hensen, E. J. M. Mechanism and Nature of Active Sites for Methanol Synthesis from CO/CO_2 on Cu/CeO_2. *ACS Catal.* 2020, *10* (19), 11532–11544. https://doi.org/10.1021/acscatal.0c02909

(69) Dang, S.; Yang, H.; Gao, P.; Wang, H.; Li, X.; Wei, W.; Sun, Y. A Review of Research Progress on Heterogeneous Catalysts for Methanol Synthesis from Carbon Dioxide Hydrogenation. *Catal. Today* 2019, *330*, 61–75. https://doi.org/10.1016/j.cattod.2018. 04.021

(70) Li, S.; Guo, L.; Ishihara, T. Hydrogenation of CO_2 to Methanol over Cu/AlCeO Catalyst. *Catal. Today* 2020, *339*, 352–361. https://doi.org/10.1016/j.cattod.2019.01.015

(71) Chang, K.; Wang, T.; Chen, J. G. Hydrogenation of CO_2 to Methanol over CuCeTiOx Catalysts. *Appl. Catal. B Environ.* 2017, *206*, 704–711. https://doi.org/10.1016/j.apcatb. 2017.01.076

(72) Wang, W.; Qu, Z.; Song, L.; Fu, Q. CO_2 Hydrogenation to Methanol over Cu/CeO_2 and Cu/ZrO_2 Catalysts: Tuning Methanol Selectivity via Metal-Support Interaction. *J. Energy Chem.* 2020, *40*, 22–30. https://doi.org/10.1016/j.jechem.2019.03.001

(73) Dasireddy, V. D. B. C.; Likozar, B. The Role of Copper Oxidation State in $Cu/ZnO/Al_2O_3$ Catalysts in CO_2 Hydrogenation and Methanol Productivity. *Renew. Energy* 2019, *140*, 452–460. https://doi.org/10.1016/j.renene.2019.03.073

(74) Ladera, R.; Pérez-Alonso, F. J.; González-Carballo, J. M.; Ojeda, M.; Rojas, S.; Fierro, J. L. G. Catalytic Valorization of CO_2 via Methanol Synthesis with Ga-Promoted Cu–ZnO–ZrO_2 Catalysts. *Appl. Catal. B Environ.* 2013, *142–143*, 241–248. https://doi.org/10.1016/j.apcatb.2013.05.019

(75) Ren, M.; Zhang, Y.; Wang, X.; Qiu, H. Catalytic Hydrogenation of CO_2 to Methanol: A Review. *Catalysts* 2022, *12* (4), 403. https://doi.org/10.3390/catal12040403

(76) Samson, K.; Sliwa, M.; Socha, R. P.; Góra-Marek, K.; Mucha, D.; Rutkowska-Zbik, D.; Paul, J.-F.; Ruggiero-Mikołajczyk, M.; Grabowski, R.; Słoczyński, J. Influence of ZrO_2 Structure and Copper Electronic State on Activity of Cu/ZrO_2 Catalysts in Methanol Synthesis from CO_2. *ACS Catal.* 2014, *4* (10), 3730–3741. https://doi.org/10.1021/cs500979c

(77) Gao, P.; Li, F.; Zhang, L.; Zhao, N.; Xiao, F.; Wei, W.; Zhong, L.; Sun, Y. Influence of Fluorine on the Performance of Fluorine-Modified Cu/Zn/Al Catalysts for CO_2 Hydrogenation to Methanol. *J. CO_2 Util.* 2013, *2*, 16–23. https://doi.org/10.1016/j.jcou.2013.06.003

(78) Xiao, S.; Zhang, Y.; Gao, P.; Zhong, L.; Li, X.; Zhang, Z.; Wang, H.; Wei, W.; Sun, Y. Highly Efficient Cu-Based Catalysts via Hydrotalcite-like Precursors for CO_2 Hydrogenation to Methanol. *Catal. Today* 2017, *281*, 327–336. https://doi.org/10.1016/j.cattod.2016.02.004

(79) Gao, P.; Zhong, L.; Zhang, L.; Wang, H.; Zhao, N.; Wei, W.; Sun, Y. Yttrium Oxide Modified $Cu/ZnO/Al_2O_3$ Catalysts via Hydrotalcite-like Precursors for CO_2 Hydrogenation to Methanol. *Catal. Sci. Technol.* 2015, *5* (9), 4365–4377. https://doi.org/10.1039/C5CY00372E

(80) Li, M. M.-J.; Zeng, Z.; Liao, F.; Hong, X.; Tsang, S. C. E. Enhanced CO_2 Hydrogenation to Methanol over CuZn Nanoalloy in Ga Modified Cu/ZnO Catalysts. *J. Catal.* 2016, *343*, 157–167. https://doi.org/10.1016/j.jcat.2016.03.020

(81) Jiang, Y.; Yang, H.; Gao, P.; Li, X.; Zhang, J.; Liu, H.; Wang, H.; Wei, W.; Sun, Y. Slurry Methanol Synthesis from CO_2 Hydrogenation over Micro-Spherical SiO_2 Support Cu/ZnO Catalysts. *J. CO_2 Util.* 2018, *26*, 642–651. https://doi.org/10.1016/j.jcou.2018.06.023

(82) Phongamwong, T.; Chantaprasertporn, U.; Witoon, T.; Numpilai, T.; Poo-Arporn, Y.; Limphirat, W.; Donphai, W.; Dittanet, P.; Chareonpanich, M.; Limtrakul, J. CO_2 Hydrogenation to Methanol over CuO–ZnO–ZrO_2–SiO_2 Catalysts: Effects of SiO_2 Contents. *Chem. Eng. J.* 2017, *316*, 692–703. https://doi.org/10.1016/j.cej.2017.02.010

(83) Dong, X.; Li, F.; Zhao, N.; Tan, Y.; Wang, J.; Xiao, F. CO_2 Hydrogenation to Methanol over Cu/Zn/Al/Zr Catalysts Prepared by Liquid Reduction. *Chin. J. Catal.* 2017, *38* (4), 717–725. https://doi.org/10.1016/S1872-2067(17)62793-1

(84) Panagiotopoulou, P. Hydrogenation of CO_2 over Supported Noble Metal Catalysts. *Appl. Catal. Gen.* 2017, *542*, 63–70. https://doi.org/10.1016/j.apcata.2017.05.026

(85) Xiong, S.; Lian, Y.; Xie, H.; Liu, B. Hydrogenation of CO_2 to Methanol over Cu/ZnCr Catalyst. *Fuel* 2019, *256*, 115975. https://doi.org/10.1016/j.fuel.2019.115975

(86) Ren, S.; Shoemaker, W. R.; Wang, X.; Shang, Z.; Klinghoffer, N.; Li, S.; Yu, M.; He, X.; White, T. A.; Liang, X. Highly Active and Selective Cu-ZnO Based Catalyst for Methanol and Dimethyl Ether Synthesis via CO_2 Hydrogenation. *Fuel* 2019, *239*, 1125–1133. https://doi.org/10.1016/j.fuel.2018.11.105

(87) Wang, W.; Tongo, D. W. K.; Song, L.; Qu, Z. Effect of Au Addition on the Catalytic Performance of CuO/CeO$_2$ Catalysts for CO$_2$ Hydrogenation to Methanol. *Top. Catal.* 2021, *64* (5), 446–455. https://doi.org/10.1007/s11244-021-01414-3

(88) Salehi, M.-S.; Askarishahi, M.; Gallucci, F.; Godini, H. R. Selective CO$_2$-Hydrogenation Using a Membrane Reactor. *Chem. Eng. Process. - Process Intensif.* 2021, *160*, 108264. https://doi.org/10.1016/j.cep.2020.108264

(89) Chen, S.; Zhang, J.; Song, F.; Zhang, Q.; Yang, G.; Zhang, M.; Wang, X.; Xie, H.; Tan, Y. Induced High Selectivity Methanol Formation during CO$_2$ Hydrogenation over a CuBr$_2$-Modified CuZnZr Catalyst. *J. Catal.* 2020, *389*, 47–59. https://doi.org/10.1016/j.jcat.2020.05.023

(90) Mureddu, M.; Lai, S.; Atzori, L.; Rombi, E.; Ferrara, F.; Pettinau, A.; Cutrufello, M. G. Ex-LDH-Based Catalysts for CO$_2$ Conversion to Methanol and Dimethyl Ether. *Catalysts* 2021, *11* (5), 615. https://doi.org/10.3390/catal11050615

(91) Pasupulety, N.; Al-Zahrani, A. A.; Daous, M. A.; Podila, S.; Driss, H. A Study on Highly Active Cu-Zn-Al-K Catalyst for CO$_2$ Hydrogenation to Methanol. *Arab. J. Chem.* 2021, *14* (2), 102951. https://doi.org/10.1016/j.arabjc.2020.102951

(92) Gao, P.; Yang, H.; Zhang, L.; Zhang, C.; Zhong, L.; Wang, H.; Wei, W.; Sun, Y. Fluorinated Cu/Zn/Al/Zr Hydrotalcites Derived Nanocatalysts for CO$_2$ Hydrogenation to Methanol. *J. CO$_2$ Util.* 2016, *16*, 32–41. https://doi.org/10.1016/j.jcou.2016.06.001

(93) Liang, Z.; Gao, P.; Tang, Z.; Lv, M.; Sun, Y. Three Dimensional Porous Cu-Zn/Al Foam Monolithic Catalyst for CO$_2$ Hydrogenation to Methanol in Microreactor. *J. CO$_2$ Util.* 2017, *21*, 191–199. https://doi.org/10.1016/j.jcou.2017.05.023

(94) Pustovarenko, A.; Dikhtiarenko, A.; Bavykina, A.; Gevers, L.; Ramírez, A.; Russkikh, A.; Telalovic, S.; Aguilar, A.; Hazemann, J.-L.; Ould-Chikh, S.; Gascon, J. Metal–Organic Framework-Derived Synthesis of Cobalt Indium Catalysts for the Hydrogenation of CO$_2$ to Methanol. *ACS Catal.* 2020, *10* (9), 5064–5076. https://doi.org/10.1021/acscatal.0c00449

(95) Shen, C.; Sun, K.; Zhang, Z.; Rui, N.; Jia, X.; Mei, D.; Liu, C. Highly Active Ir/In2O3 Catalysts for Selective Hydrogenation of CO$_2$ to Methanol: Experimental and Theoretical Studies. *ACS Catal.* 2021, *11* (7), 4036–4046. https://doi.org/10.1021/acscatal.0c05628

(96) Lu, Z.; Sun, K.; Wang, J.; Zhang, Z.; Liu, C. A Highly Active Au/In$_2$O$_3$-ZrO$_2$ Catalyst for Selective Hydrogenation of CO$_2$ to Methanol. *Catalysts* 2020, *10* (11), 1360. https://doi.org/10.3390/catal10111360

(97) Wang, J.; Sun, K.; Jia, X.; Liu, C. CO$_2$ Hydrogenation to Methanol over Rh/In2O3 Catalyst. *Catal. Today* 2021, *365*, 341–347. https://doi.org/10.1016/j.cattod.2020.05.020

(98) Gothe, M. L.; Pérez-Sanz, F. J.; Braga, A. H.; Borges, L. R.; Abreu, T. F.; Bazito, R. C.; Gonçalves, R. V.; Rossi, L. M.; Vidinha, P. Selective CO$_2$ Hydrogenation into Methanol in a Supercritical Flow Process. *J. CO$_2$ Util.* 2020, *40*, 101195. https://doi.org/10.1016/j.jcou.2020.101195

(99) Wang, Y.; Wu, D.; Liu, T.; Liu, G.; Hong, X. Fabrication of PdZn Alloy Catalysts Supported on ZnFe Composite Oxide for CO$_2$ Hydrogenation to Methanol. *J. Colloid Interface Sci.* 2021, *597*, 260–268. https://doi.org/10.1016/j.jcis.2021.03.135

(100) Hu, J.; Yu, L.; Deng, J.; Wang, Y.; Cheng, K.; Ma, C.; Zhang, Q.; Wen, W.; Yu, S.; Pan, Y.; Yang, J.; Ma, H.; Qi, F.; Wang, Y.; Zheng, Y.; Chen, M.; Huang, R.; Zhang, S.; Zhao, Z.; Mao, J.; Meng, X.; Ji, Q.; Hou, G.; Han, X.; Bao, X.; Wang, Y.; Deng, D. Sulfur Vacancy-Rich MoS$_2$ as a Catalyst for the Hydrogenation of CO$_2$ to Methanol. *Nat. Catal.* 2021, *4* (3), 242–250. https://doi.org/10.1038/s41929-021-00584-3

(101) Bahruji, H.; Bowker, M.; Hutchings, G.; Dimitratos, N.; Wells, P.; Gibson, E.; Jones, W.; Brookes, C.; Morgan, D.; Lalev, G. Pd/ZnO Catalysts for Direct CO$_2$ Hydrogenation to Methanol. *J. Catal.* 2016, *343*, 133–146. https://doi.org/10.1016/j.jcat.2016.03.017

(102) Bahruji, H.; Bowker, M.; Jones, W.; Hayward, J.; Esquius, J. R.; Morgan, D. J.; Hutchings, G. J. PdZn Catalysts for CO$_2$ Hydrogenation to Methanol Using Chemical Vapour Impregnation (CVI). *Faraday Discuss.* 2017, *197* (0), 309–324. https://doi.org/10.1039/C6FD00189K

(103) Qu, J.; Zhou, X.; Xu, F.; Gong, X.-Q.; Tsang, S. C. E. Shape Effect of Pd-Promoted Ga$_2$O$_3$ Nanocatalysts for Methanol Synthesis by CO$_2$ Hydrogenation. *J. Phys. Chem. C* 2014, *118* (42), 24452–24466. https://doi.org/10.1021/jp5063379

9 Synthetic Gasoline

José Gonzalo Hernández-Cortez
Mexican Petroleum Institute, México City, México

9.1 INTRODUCTION

At present, the transport sector is responsible for the consumption of almost 40% of the world's energy, and it does so primarily uses liquid fuels derived from fossil resources. Furthermore, the demand for such fuels is predicted to increase significantly from 99.1 million barrels in 2022 to 121.5 million barrels in 2050.[1] Environmental regulations and the growing energy crisis also play an important role. The conversion of synthesis gas, derived from coal, biomass, and natural gas, into cleaner fuels is of great interest worldwide.[2,3] Against this background, the question arises: is it possible to maintain such consumption with the current oil reserves?

This chapter provides an overview of two current catalytic processes associated with CO hydrogenation to produce clean liquid fuels considering their relevance in the sustainable energy scene and their importance at a global level. These processes can be divided into (i) obtaining hydrocarbons and liquid fuels by Fischer–Tropsch Synthesis (FTS) and (ii) the transformation of methanol to hydrocarbons of the gasoline range (methanol to gasoline, MTG). Both are catalytic processes in which the important step is the activation of the CO molecule for FTS and the CH_3OH molecule for MTG, dissociating them completely for the synthesis of hydrocarbons.[4] As a whole, the process of synthesis of hydrocarbons in the gasoline range requires obtaining a raw material called synthesis gas or syngas,[5,6] a mixture of H_2 and CO, the relative concentration of which depends on the carbon precursor used. These sources are coal, natural gas, and biomass; the process takes its name from the origin of the synthesis gas and is referred to as coal-to-liquids (CTL), gas-to-liquids (GTL), and biomass-to-liquids (BTL) processes[2,7,8] as shown in Figure 9.1. Special attention is devoted to distinct factors related to i) knowledge of the FTS and MTG processes in general, ii) an overview of the reaction mechanism, and iii) current development of the catalysts used.

Catalysis is a key player in the growing development of society. Processes such as the production of clean fuels, the production of hydrogen, the removal of pollutants from water and air, the selective synthesis of pharmaceuticals and high value-added products, as well as the production of polymers, resins, fibers, and lubricants, among thousands of others, involve at least one catalytic step. It is estimated that 85–90% of products in the industry involve at least one catalytic process in production.[9]

In many catalytic processes, especially those related to the energy sector, heterogeneous catalysts are used, consisting of high-area solids on which metallic particles

DOI: 10.1201/9781003517283-9

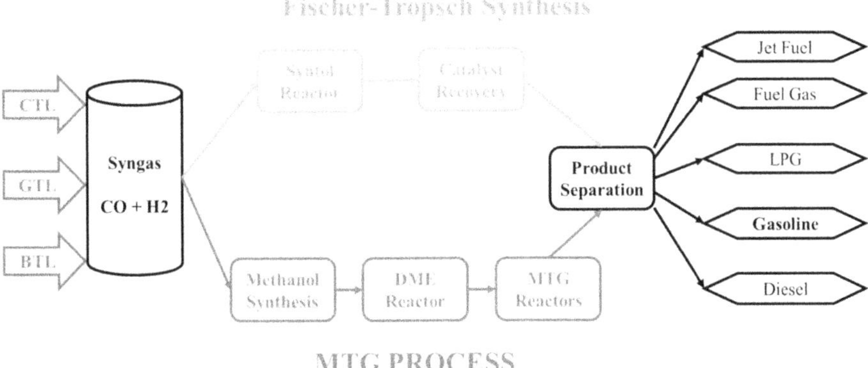

FIGURE 9.1 A simplified diagram of the coal-to-liquids (CTL), gas-to-liquids (GTL), and biomass-to-liquids (BTL) processes.

are deposited. Typical examples of the use of heterogeneous catalysts at the industrial level include three-way catalysts for the removal of pollutants resulting from the use of gasoline in internal combustion motors, processes related to the transformation of CO, the production of hydrogen from methane reforming, the water gas shift reaction for hydrogen purification, the synthesis of ammonia and its derivatives, etc.

The important events that occur in heterogeneous gas-solid catalytic processes are (i) external diffusion of reactants into the layer surrounding the solid; (ii) internal diffusion of reactants into the porous solid; (iii) adsorption of reactants; (iv) catalytic reaction; (v) desorption of products; (vi) internal diffusion of products; and (vii) external diffusion of reaction products.[10–12]

9.2 FISCHER–TROPSCH PROCESS TO GENERATE LIQUID FUELS

Syngas can be applied to directly produce clean and environmentally friendly liquid fuels such as gasoline (C_5–C_{12}), jet fuel (C_8–C_{16}), and diesel (C_{10}–C_{20}), for which the FT process is applied. A plant with this synthesis process relies on the introduction of acid catalysts to tailor the carbon chains, thereby increasing the proportion of hydrocarbons in the gasoline range.[13–15] Some researchers have focused on the structures and properties of the catalysts by introducing supports rich in acid sites and adding metal promoters to enhance the hydrocracking and isomerization reactions.

The process of producing hydrocarbons by CO hydrogenation was first described by Franz Fischer and Hans Tropsch in the 1920s.[16] Indeed, the main application of synthesis gas from coal is the production of synthetic hydrocarbons for transport fuels by the FT process. The production of synthesis gas from coal gasification continues to grow worldwide at a rate of about 10% per year, which means that the technology remains of significant industrial importance.[8] This was mainly done in South Africa by the SASOL company and was also one of the methods used by the Germans in World War II to generate liquid fuels; in fact, direct liquefaction was the main method used to produce liquid fuels in Germany in the 1940s.[16,17] However, it is not the only liquid gasification process. The FT synthesis reaction can be presented in general form by Equation 9.1.

$$CO + n\,H_2 \rightarrow (\,-CH_2-\,)_x + H_2O \tag{9.1}$$

The FT process transforms carbon atoms and forms alkanes up to a range of at least 20 carbon atoms. In reality, it is a polymerization process and follows the Anderson–Schulz–Flory (ASF) distribution and posits an upper limit on the theoretical selectivity of the FT process for gasoline (C_5–C_{12}) at ~48 wt.% in conventional reactors.[18,19] No single pure alkane will be obtained from the FT process and there will be a distribution of products. As with all chemical reactions, it will be necessary to adjust the reaction variables, such as temperature, pressure, residence time, and the addition of a catalyst. With a suitable selection of reaction variables, in principle, it is possible to obtain from methane to high-molecular-weight waxes. The intention is to maximize the production of liquid fuel for transport.

The liquids obtained by the FT process are exceptionally clean fuels, the product is almost zero sulfur, low in aromatic compounds, and is composed of straight-chain alkanes. The FT gasoline that comes directly from the reaction is not large as it has a low octane number. Recall that branched alkanes and aromatic compounds have higher octane ratings. Since the products of the FT process tend to be straight-chain alkanes, an isomerization reaction is required and an appropriate catalyst must be used for catalytic reforming. Recently, composite or bifunctional catalysts containing metal oxide components and zeolites have been applied to the FT process for the conversion of syngas to olefins, aromatics, and fuels.[4,20–23]

9.2.1 General Aspects of Fischer–Tropsch Synthesis

The main reaction of the FT process is the production of paraffins and olefins (Equations 9.2 and 9.3). However, secondary reactions are also conducted, including the synthesis of oxygenated products (Equation 9.4), the water-gas-shift reaction (Equation 9.5), and the Boudouard reaction (Equation 9.6).

$$n\,CO + (2n + 1)\,H_2 \rightarrow C_nH_{2n+2} + n\,H_2O \text{ (paraffins)} \tag{9.2}$$

$$n\,CO + 2n\,H_2 \rightarrow C_nH_{2n} + n\,H_2O \text{ (olefins)} \tag{9.3}$$

$$n\,CO + 2n\,H_2 \rightarrow C_nH_{2n+2}O + (n-1)\,H_2O \text{ (alcohols)} \tag{9.4}$$

$$CO + H_2O \rightarrow CO_2 + H_2 \text{ (water} - \text{gas} - \text{shift)} \tag{9.5}$$

$$2\,CO \rightarrow C + CO_2 \text{ (Boudouard)} \tag{9.6}$$

Although the details of the mechanism of FT synthesis are not fully understood, most researchers accept that hydrocarbon formation occurs through a polymerization reaction of CHx* units formed in situ on the catalyst surface from CO and H_2 and which function as initiator and monomer species in the chain growth process (Figure 9.2). There are several possibilities for chain growth termination: (i) hydrogenation to form

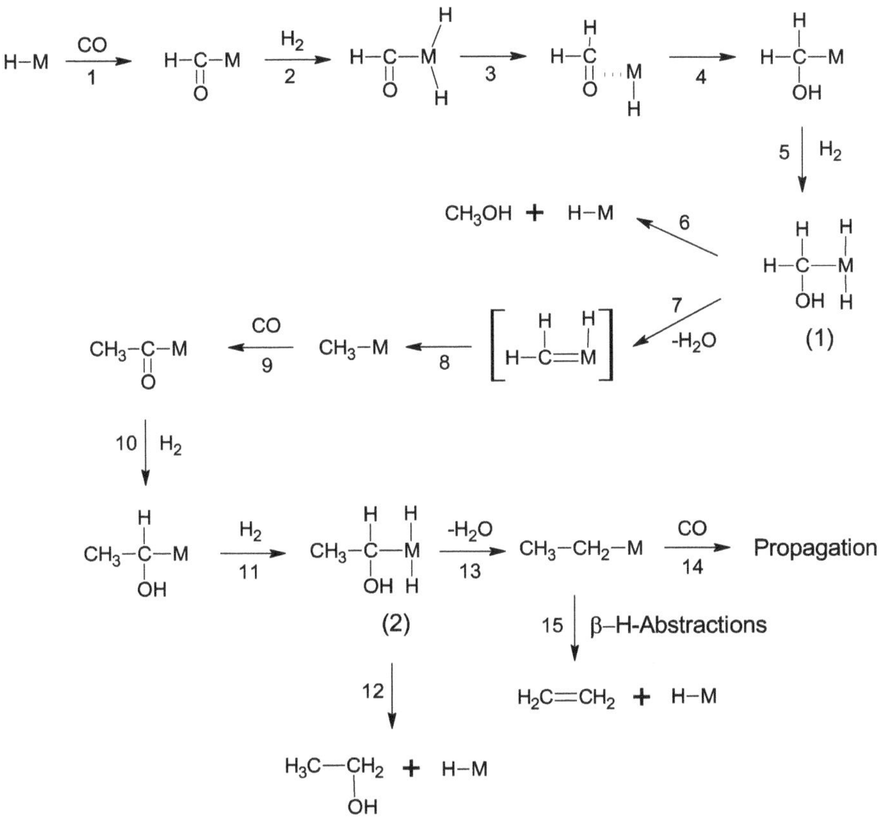

FIGURE 9.2 Mechanism of the Fischer–Tropsch reaction.

paraffins; (ii) dehydrogenation to form preferentially 1-olefins; and (iii) CO insertion and subsequent hydrogenation to generate oxygenated compounds.

The general mechanism of FT synthesis is depicted in Figure 9.2:

(Step 1)   Indicates that the CO coordinates with the active center and inserts into the M-H bond.

(Step 2)   A hydrogen molecule is added to the metal center.

(Step 3)   Reductive elimination of the acyl ligand and one of the hydride ligands give formaldehyde and one metal hydride; the former, however, does not leave the metal center but remains coordinatively bound via its C=O group (aldehydes are not primary products of the FT reaction).

(Step 4)   Addition of the metal hydride to the aldehyde.

(Step 5)   Followed by another oxidative addition of H_2 forming intermediate (I) which can react in two ways.

(Step 6)   Reductive elimination to give methanol and a metal hydride that can continue the kinetic chain or

(Step 7)   Removal of water with intermediate formation of a carbenoid ligand.

(Step 8)   Rearrangement of the carbenoid ligand to give a bound methyl ligand.

(Step 9)   The next CO molecule can now be coordinated and inserted.

(Step 10)   Summarizes the results of three steps, corresponding to steps 2 to 4, with the difference that the carbon chain now increases by one unit.

(Step 11)   The oxidative addition of H_2 leads to the intermediate (2) which again can undergo two alternative reactions.

(Step 12)   An alternative reaction yielding the alcohol by reductive elimination or

(Step 13)   The alkyl group by H_2O elimination, which summarizes the two steps corresponding to steps 7 and 8.

(Step 14)   The alkyl metal compound can add CO and thus contribute to the chain propagation or

(Step 15) β-H transfer giving an α-olefin (ethylene at this stage) and a metal hydride that continues the kinetic chain.

9.2.2 Catalysts Used in FTS

In general, the catalysts are based on metals such as Co and Ru, which are highly active in CO dissociation and are therefore ideal candidates for hydrocarbon synthesis. In recent years, the use of bifunctional catalysts combining acid zeolites with active components has become a promising means to achieve FT synthesis. Long-chain linear carbon hydrocarbons are the main products of traditional FT synthesis, but acid zeolites can facilitate secondary reactions of long-chain linear carbon hydrocarbons formed on metal active sites.[24–26] The secondary reactions include the hydrocracking of long-carbon-chain hydrocarbons to low-carbon-chain hydrocarbons, the alkylation of olefins to alkanes, and the isomerization of linear hydrocarbons to branched hydrocarbons. The special porous structure of some zeolites can limit the mass transfer of long-chain hydrocarbons leaving the zeolite channel into the gas phase and adequate acidity can enhance secondary reactions. These properties are beneficial for the selective production of gasoline-type liquid fuels with more branched alkanes. For example, the addition of zeolite (ZSM-5) increases the number of acid sites of a catalyst, effectively reducing the selectivity towards heavy hydrocarbons.[27] On the other hand, alkaline treatment of a zeolite (ZSM-5) produces a material with both microporous and mesoporous structures, increasing the dispersion of the active metal and soft acid sites, leading to better selectivity for gasoline.[28–30] Some recently used catalysts for the production of gasoline in the FT reaction are listed in Table 9.1.

Reports on Co/Y-β zeolite and Co/Ce-β zeolite catalysts demonstrated improved catalytic activity and stability. These ions, Y^{3+} and Ce^{3+}, enhance the strength of the CO–active site (metal) bond while weakening the CO bond, which promotes increased selectivity in obtaining gasoline (C_5–C_{12}). In addition, doping with Y^{3+} promoted the dispersion of Co particles, and the addition of Ce^{3+} strengthened the reducibility facilitating the conversion of CO. The acidity of the β-zeolite also played an important role in the reaction since the use of La^{3+} resulted in light products (C_2–C_4).[32] The acidity of zeolites is beneficial for the adjustment of product distributions in the gasoline range through secondary reactions of cracking, isomerization, and aromatization.[38] Furthermore, the acidity of the zeolite has a strong impact on the formation and dispersion of active Co^0 species, as well as on CO conversion and product distribution. Cerium can store oxygen, and the dissociation of CO molecules in the metal Co will

TABLE 9.1

Catalysts Used to Produce Gasoline by the FT Reaction

Catalyst	H_2/CO	P, MPa	T, °C	WHSV or GHSV, h^{-1}	CO,% Conv.	Gasoline C_5–C_{12}, %	Ref.
Co/H-β[a]	2	2	260	1064	90.5	43.5	[31]
Co/Ce-β[a]	2	2	260	1064	91.7	45.5	[31]
Co/Y-β[a]	2	2	260	1064	93.4	43.7	[31]
ZSM-5/Co-Al$_2$O$_3$/M[b]	2	1.2	230	–	78.7	93.3	[32]
Co-Al$_2$O$_3$/M[b]	2	1.2	230	–	81.7	63.3	[32]
Co/(Z5+S15)[c]	2	2	240	1000	85.0	62.0	[33]
Co-HZSM-5	2	2	235	3000	74.0	76.0	[34]
Ru/meso-β-0.10M[d]	1	2	260	–	~30.0	75.0	[14]
Ru/meso-β-0.15M[d]	1	2	260	–	~32.0	77.0	[14]
Co/SZ-A[e]	2	2	270	6750*	92.0	80.4	[35]
Co/SZ-B[e]	2	2	270	6750*	88.1	89.5	[35]
Zn$_2$Mn$_1$Ox/SAPO-11	1	4	360	1000	20.3	76.7	[36]
Zn$_2$Mn$_1$Ox/ZSM-22	1	4	360	1000	19.3	64.2	[36]
20CsDP-Co/Al$_2$O$_3$[f]	2	2	250	6000	85.3	48.8	[37]
20HDP-Co/Al$_2$O$_3$[g]	2	2	250	6000	82.3	47.7	[37]

[a] β-zeolite.

[b] Cordierite monolith substrates (2MgO:2Al$_2$O$_3$:5SiO$_2$, Corning, 200 psi, L: 7.5 cm).

[c] SBA-15 = S15 and Z5 = ZSM-5.

[d] β-zeolites with simple alkaline post-treatment (0.10 and 0.15 M of NaOH).

[e] Co/SZA and Co/SZ-B are bifunctional catalysts produced by coupling Co/SBA-15 and HZSM-5 with different proximities between active Co^0 sites and acidic sites.

[f] CsDP is $Cs_{5.5}H_{0.5}P_2W_{18}O_{62}$.

[g] HDP is $H_6P_2W_{18}O_{62}$.

* mL/(g h).

form O* species which will react with CO to form CO_2.[39] When Ce is present, the O* species react with Ce and are not available for reaction with CO to form CO_2.

Another research group focused on exploring the hypothesis that the combination of enhanced process efficiency using cordierite monoliths (2MgO:2Al$_2$O$_3$:5SiO$_2$, Corning) with the isomerization and cracking capability of ZSM-5 can improve selectivity. Monolithic catalysts can operate with low-pressure drops, high geometric surfaces, high mass transfer coefficients, and short diffusion lengths, thus relaxing the mass and heat transfer limitations of spherical and pellet catalysts. Co catalysts supported on monoliths coated with ZSM-5 showed high selectivity to gasoline (C_5–C_{12}). The addition of ZSM-5 to the monolithic catalyst not only improved gasoline selectivity but also its quality.[32] This is due to improved diffusion limitation in lower mesopore volumes and hydrocracking of primary hydrocarbons in Brönsted-type acid sites (BAS).[40]

It has been reported that the middle acid sites of the physically mixed Co/S15 + Z5 catalyst facilitated the consecutive hydrocracking of the primary hydrocarbons formed on Co/SBA-15; consequently, the selectivity of the C_5–C_{11} hydrocarbons improved from 51% to 62%. The Co/(S15 + Z5) catalyst exhibited high catalytic stability after being used at 240 °C for 120 h due to the synergistic effect of hydrogenolysis and cracking, which suppressed coke deposition.[33]

Other research groups worked on a bifunctional Co/SZ-B catalyst by combining Co/SBA-15 and HZSM-5. The evaluated catalyst exhibited high selectivity to gasoline distillate (C_5–C_{11}) due to carbon chain growth on Co^0 active sites and hydrocracking on Co^0 active sites and acid sites, as well as isomerization on acid sites. The Co/SZ-B catalyst, with a micrometer-scale distance between the Co^0 active sites and the acid sites, was found to exhibit excellent selectivity to gasoline of up to 89.54% in the liquid product.[35]

Cheng et al. considered that the mesoporosity and unique acidity of β-zeolite contribute to the selective hydrocracking of heavier hydrocarbons to C_5–C_{11} hydrocarbons. On this basis, they prepared a series of mesoporous β-zeolites by simple alkaline post-treatment. The modification of mesopore size and volume was a function of the NaOH concentration used. The higher NaOH concentration generated mesopores with larger sizes and volumes because the crystalline structure of the β-zeolite underwent a partial collapse. Finally, they concluded that the use of meso-β-zeolite, with Lewis-type acidity domain, as a support not only resulted in decreased CH_4 selectivity but also significantly reduced C_{12+} selectivity. In addition, a selectivity of 77% toward C_5–C_{11} hydrocarbons was achieved with an isoparaffin to n-paraffin ratio of 2.7 over a Ru/meso-β-0.15 M NaOH catalyst.[14]

Recent studies on the use of zeolites, such as SAPO-11 and ZSM-22 with 1D 10-MR channel type, were employed with the bifunctional oxide-zeolite (OX-ZEO) catalyst concept, it was reported that high-quality gasoline can be synthesized directly from syngas by combining $Zn_aMn_bO_x$ with this type of zeolites. In addition, the overall CO conversion activity can be tuned by the composition and structure of the metal oxides. In particular, the presence of Mn-doped ZnO can boost CO conversion. As a result, the Zn_2Mn_1Ox-SAPO-11 composite catalyst provides high activity and selectivity towards C_5–C_{11} of 76.7% with only 2.3% CH_4.[36]

Finally, studies were reported using polyoxometalates (POM), which possess a unique strong Brønsted-type acidity. The Wells–Dawson type POM salt, $Cs_{5.5}H_{0.5}P_2W_{18}O_{62}$ (CsDP), combined with the conventional FT catalyst, Co/Al_2O_3, formed a highly efficient bifunctional catalyst for the direct production of clean gasoline from syngas. The C_5–C_{12} selectivity approached 50% with a 129% increase in gasoline selectivity. This indicated that the acidity properties and the type of POM structure influence the hydrocracking of hydrocarbons and finally impact the gasoline yield.[37]

9.3   GASOLINE FROM METHANOL (MTG)

Conventional methods to produce gasoline are generally the distillation of crude oil and processes such as alkylation and catalytic cracking, although the coal-to-liquid process is carried out in some countries with coal resources. Liquid fuels, and gasoline, are the main sources of energy for transport. Their worldwide demand is expected to increase, their use has clear environmental concerns and crude oil reserves are in decline,[41] making alternative sources of energy of great interest. On this basis, the

MTG process[42] is presented as an efficient technology for the production of high-octane gasoline at a lower cost; methanol can be produced from a wide range of resources, such as coal, natural gas, or biomass, that is, from renewable sources.[3,43,44]

This MTG process emerged in the early 1970s at *ExxonMobil*,[45] which involves the transformation of a methanol stream into gasoline when it is brought into contact with a ZSM-5 zeolite,[46] which has a high specific surface area, high acidity, and well-defined structure. However, the HZSM-5 catalyst suffers from low selectivity toward C_5–C_{10} alkanes, high yields of aromatics, and carbon deposition, thus requiring frequent regeneration. High aromatic contents in gasoline are not desired.[47]

The major products of the MTG process are gasoline (85%) and LPG (easily condensable gases, that is, propane, n-butane, and isobutane) in a proportion of 13–14%. Nevertheless, the concentration of aromatics and the nonlinear nature of the paraffins give the final product a high quality as gasoline; the only difference with that obtained from crude oil is the presence of durene (1,2,4,5-tetramethyl-benzene).[48–50] ExxonMobil established a durene limit value of 2 wt % in final gasoline for use in automobiles. The process is economically viable when oil prices are high.[49]

The MTG process uses a special zeolite catalyst with a pore size such that molecules up to C_{10} can exit the catalyst.[46] The ZSM-5 zeolite with a three-dimensional interconnected channel system, high external surface area, and excellent shape-selective effects is the most suitable catalyst for the MTG reaction.[50] With these catalyst characteristics, larger molecules cannot be generated; consequently, a product with no carbon molecules larger than C_{10} is produced. This process produces aromatics and branched-chain alkanes, which means that every high-octane gasoline is obtained.[51]

The distribution of hydrocarbons obtained in methanol transformation could be divided into three types, that is, methanol-to-olefins (MTO), MTG, and methanol-to-propylene (MTP) (see Figure 9.3). MTG's first commercial operation was in New

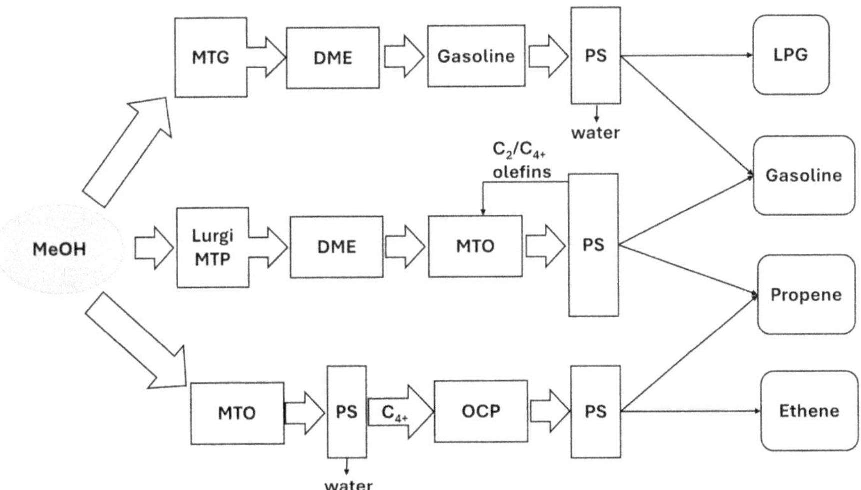

FIGURE 9.3 A general outline of the processes: methanol-to-gasoline (MTG), methanol-to-olefins (MTO), and methanol-to-propylene (MTP). DME= dimethyl ether, LPG= liquified petroleum gas, OCP= olefin cracking process, PS= product separation.

Zealand in 1985.[52,53] In addition, MTO process technologies, including Hydro/UOP's MTO[54] and Lurgi's MTO (MTP),[55] have also achieved commercial operation.

9.3.1 General Mechanism of the MTG Process

The basic mechanism recognized for the MTG process corresponds to an equilibrium, where methanol is first dehydrated and dimethyl ether (DME) is produced. The resulting mixture of methanol, water, and DME is then dehydrated again and the C–C bond is formed, giving light olefins.[56,57] In the last stage of the reaction, light olefins are converted into paraffins, aromatics, naphthalene, and heavier olefins.[58,59] The selectivity of this last stage is determined by the conditions under which the reaction takes place.[48,49] The following mechanism reaction is commonly assumed:[60]

$$2\ CH_3OH \leftrightarrow CH_3OCH_3 + H_2O \text{ (dimethyl ether)} \tag{9.7}$$

$$CH_3OCH_3 \rightarrow -\ C_nH_{2n} - +\ H_2O \text{ (olefins)} \tag{9.8}$$

$$-C_nH_{2n} - \rightarrow \text{Hydrocarbons} \tag{9.9}$$

$$-C_nH_{2n} - +\ \text{Hydrocarbons} \rightarrow \text{Gasoline range hydrocarbons} \tag{9.10}$$

As an important reaction in heterogeneous catalysis, MTG conversion provides a sustainable way for the production of valuable platform chemicals such as olefins, gasoline, and aromatics. Although many commercial plants have been operated recently, there is a need to improve catalytic efficiency, that is, selectivity to the desired product and catalyst lifetime. As an acid-catalyzed reaction, methanol is first dehydrated to DME over BAS. Subsequently, the mixture of methanol and DME is transformed into hydrocarbons by isomerization, alkylation, and cracking reactions, which depend on the type of acid sites (Brönsted or Lewis) and acid strength (strong, medium, or weak). Based on the reaction mechanism, the design, and the optimization of the catalyst to improve the selectivity of the target product and promote the lifetime of the catalyst are also highlighted.

9.3.2 Catalysts Used in MTG

Zeolite-based catalysts are highly valued in MTG reactions because of their exceptional stability, selectivity towards target products, and environmental compatibility.[61] Among zeolite sieves, ZSM-5 and ZSM-11 have shown remarkable catalytic efficiency for MTG reactions. Nevertheless, their 3D channels and 10-membered ring (MR) pore structure tend to produce aromatic hydrocarbon-rich gasoline.[46,62] For example, Jia et al. evaluated the catalytic performance of zeolite ZSM-5 at 400°C, 0.1 MPa, and a WHSV of 1.2 h^{-1}, achieving 100% methanol conversion and 36.56% selectivity for light aromatic hydrocarbons during a 16 h reaction period.[63] Although aromatic hydrocarbons can increase the octane rating of gasoline, their presence in gasoline can also increase the release of harmful gases from motor vehicles, which are detrimental to both human health and the ecological environment.[64]

TABLE 9.2
Catalysts Used in the MTG Process

Catalyst	P, MPa	T, °C	WHSV or GHSV, h^{-1}	MeOH, % Conv.	Gasoline Yield, %	Ref.
ZSM-5, SiO_2/Al_2O_3 = 217	1	375	2	100.0	70.0	[65]
ZSM-5, SiO_2/Al_2O_3 = 23	1	375	2	94.0	~50.0	[65]
ZB0[a]	0.1	400	20	80.0	12.5	[66]
ZB30[a]	0.1	400	20	87.0	22.0	[66]
2%CuO/ZSM-5	–	425	–	94.6[b]	83.0	[67]
4%CuO/ZSM-5	–	425	–	93.3[b]	88.3	[67]
ZSM-5/MCM-48[c]	1	360	2	97.0	33.6	[68]
ZSM-5/MCM-48[c]	1	420	2	99.8	31.7	[68]
Z5-1.0[d]	0.1013	400	10	100.0	37.1	[69]
ZS-0.4[d]	0.1013	400	10	100.0	34.4	[69]
H[Fe,Al]ZSM-5 (0.4)[e]	0.1	390	4.0	99.0	30.5	[70]
H[Fe,Al]ZSM-5 (0.8)[e]	0.1	390	4.0	99.0	32.7	[70]
ZM-0.16-25[f]	0.1	400	10	99.9	29.3	[71]
HZ-Boeh[g]		450	11.3[h]	99.9	28.8	[72]
HZ-Bent[g]		450	11.3[h]	99.9	26.5	[72]
HZ-Bent+IE[g]		450	11.3[h]	99.9	30.3	[72]

[a] ZB(x), where x is the wt % of β-zeolite in the synthesis gel (X:0, 30) in relation to the silica source.

[b] Using (70% water/30% methanol feed).

[c] ZSM-5/MCM-48 aluminosilicate composite material was synthesized by a simple two-step crystallization process.

[d] ZSM-5 (SiO_2/Al_2O_3 = 50) treatment with (TBAOH)/NaOH mixture having different mole ratios.

[e] Nanocrystalline H[Fe, Al]ZSM-5 zeolites with different Fe content.

[f] Synthesis of ZSM-5 by the seed-induced method, $0.16Na_2O\text{-}1SiO_2\text{-}0.02Al_2O_3\text{-}25H_2O\text{-}0.01$ Seed.

[g] HZ-Boeh (50wt% HZSM-5 + 30 wt% Boehmite + 20wt% Al_2O_3), HZ-Bent (25wt% HZSM-5 + 30 wt% Bentonite + 45wt% Al_2O_3), and HZ-Bent+IE (25wt% HZSM-5 + 30 wt% Bentonite + 45wt% Al_2O_3) + ion exchange treatment with a 0.6 N solution of HCl.

[h] W/F_{AO} ($g_{cat}\cdot h\cdot mol_{MeOH}^{-1}$).

The design of new micro/mesopore composite materials will moderate pore size distribution and reduce surface acidity, thereby improving selectivity towards C_5–C_{10} alkanes and reducing aromatic formation and carbon deposition. Catalyst deactivation due to coke formation, a common operational problem, can be eliminated by in situ combustion, thus achieving stable and continuous operation. Significantly, the influential co-catalytic role of oxymethylene species is related to the formation of gasoline, which affects the MTG process more than carbonyl species. With these premises, some catalysts recently used in the MTG reaction are shown in Table 9.2.

ZSM-5 nanocrystal catalysts with different SiO_2/Al_2O_3 ratios were reported. These catalysts with similar crystal sizes and structural properties showed an effect on the MTG reaction; it was reported that increasing the SiO_2/Al_2O_3 ratio reduced the durene yield and improved the durability of the catalyst. The sample with a SiO_2/Al_2O_3 ratio of 217 showed the highest methanol conversion, high gasoline yield, and very low coke formation.[65]

The dual catalyst ZSM-5/Beta has been synthesized by the hydrothermal method and used in the MTG process. The authors concluded that the incorporation of Beta zeolite into the ZSM-5 structure significantly improved the activity of the catalyst. The ZB30 catalyst had the highest total surface area, the highest external surface area, and the highest acidity; when evaluated in the MTG reaction, the catalyst lifetime was prolonged by 65%, the coking rate was decreased, and the selectivity of alkylated aromatics was improved compared to ZSM-5. The results showed that pore structure and acidity caused synergistic effects, which effectively determined the performance of the catalyst in the MTG reaction.[66] The main products of Beta zeolite are light olefins like propene and some higher alkenes.[73,74]

Other authors reported that the CuO-modified zeolite exhibited a remarkable increase in selectivity towards aromatics and higher hydrocarbons over the ZSM-5 catalyst. The 4%CuO/ZSM-5 catalyst resulted in higher activity for the conversion of MTG-range hydrocarbons, especially aromatics.[67]

A ZSM-5/MCM-48 aluminosilicate composite material was reported as a catalyst, which was synthesized by a simple two-step crystallization process. The catalyst was shown to possess high activity and stability with low formation of aromatic hydrocarbons in the conversion of MTG. The high activity and stability of the ZSM-5/MCM-48 compound catalyst were attributed to the improved mass transfer properties and reduced diffusion limitations due to better pore size distribution and lower pore surface acidity.[68] Interconnected microporous and mesoporous channel systems have been shown to greatly influence the distribution of hydrocarbon products compared to pure ZSM-5 zeolite.

The influence of chemical treatment of ZSM-5 ($SiO_2/Al_2O_3 = 50$) with a tetrabutylamine hydroxide/NaOH mixture was reported. By increasing the TBA^+/OH^- ratio, the crystalline properties, the surface areas, the micropore volumes, and the number of strong acid sites (BAS) were improved and mesopore volumes decreased. The treatment with pure tetrabutylamine hydroxide (TBAOH) ensured the formation of a tight and uniform intracrystalline mesoporosity and large amounts of strong acid sites (Brönsted type) in the zeolite, which contribute to the higher yield of liquid hydrocarbons in the MTG reaction.[69]

H[Fe, Al]ZSM-5 nanocrystalline zeolites with different Fe contents were synthesized by the hydrothermal method. The results indicated that Fe species could be incorporated into the ZSM-5 zeolite structure in the form of isolated Fe^{3+} and small oligomeric FexOy clusters, resulting in the formation of some new active sites and a lower content of BAS. The incorporation of Fe improved the activity and stability of the nanocrystalline structure, and the methanol conversion over the FNZ-0.4 and FNZ-0.8 catalysts was maintained at more than 99% for about 110 h. These catalysts were the most active and stable during the same reaction time, and the gasoline yields reached up to 30.5% and 32.7%, respectively. New active sites were created on the surface of the zeolites and the ratios of B/L acid sites decreased rapidly. In addition, the crystal size of H [Fe, Al] ZSM-5 was smaller than that of unmodified ZSM-5, leading to a lower concentration of aromatics and a higher isoparaffin content in the gasoline products.[70]

A mixture of zeolite HZSM-5 (as active material), boehmite and bentonite (as the binder), and alumina (as inert filler) were used to prepare different catalysts. The product distribution was very similar in all cases, with the yield of light hydrocarbons always being higher than that of gasoline. Among the catalysts evaluated, the catalyst containing boehmite as a binder (HZ-Boeh) proved to be the most suitable due to its high mechanical strength, high aromatic yield, and lower durene yield. The catalyst contains boehmite as a binder, with a composition of 50 wt. % of HZSM-5, 30 wt. % of boehmite, and 20 wt. % of alumina as the most suitable catalyst. It showed the highest ratio of liquids to gases did not have the fluidization problems encountered with the bentonite catalysts and had the lowest selectivity to durene.[72]

9.4 CONCLUSION AND PERSPECTIVES

The production of cleaner fuels, which do not generate environmental emissions, mainly CO_2, is a major challenge. This has led to great challenges in research to avoid carbon emissions. Gasoline (C_5–C_{11}) is essential for transport and petrochemical industries, produced mainly through the cracking and refining of oil, is energy intensive, and lacks product selectivity. With the growing demand for gasoline, there is an urgent need to explore alternative synthetic routes using nonpetroleum resources. Gasoline production by FT and MTG processes has the potential to produce renewable gasoline, a crucial solution to mitigate environmentally harmful emissions (e.g., carbon dioxide and sulfur dioxide) associated with the use of coal, oil, and natural gas. These FT and MTG technologies facilitate the production of green fuel when the feedstock (syngas and methanol) comes from renewable sources, such as biomass, the fourth most important energy resource in the biosphere. This chapter provides only a small overview of the research that has been conducted and the technologies that exist in this respect. The reader is encouraged to search further for information in the field.

REFERENCES

(1) *International Energy Outlook - U.S. Energy Information Administration (EIA)*. https://www.eia.gov/outlooks/ieo/data.php (accessed April 20, 2024)

(2) Shafer, W. D.; Gnanamani, M. K.; Graham, U. M.; Yang, J.; Masuku, C. M.; Jacobs, G.; Davis, B. H. Fischer–Tropsch: Product Selectivity–The Fingerprint of Synthetic Fuels. *Catalysts* 2019, *9* (3), 259. https://doi.org/10.3390/catal9030259

(3) Iglesias Gonzalez, M.; Kraushaar-Czarnetzki, B.; Schaub, G. Process Comparison of Biomass-to-Liquid (BtL) Routes Fischer–Tropsch Synthesis and Methanol to Gasoline. *Biomass Conv. Bioref.* 2011, *1* (4), 229–243. https://doi.org/10.1007/s13399-011-0022-2

(4) Zhang, P.; Tan, L.; Yang, G.; Tsubaki, N. One-Pass Selective Conversion of Syngas to Para-Xylene. *Chemical Science* 2017, *8* (12), 7941–7946. https://doi.org/10.1039/C7SC03427J

(5) Wilhelm, D. J.; Simbeck, D. R.; Karp, A. D.; Dickenson, R. L. Syngas Production for Gas-to-Liquids Applications: Technologies, Issues and Outlook. *Fuel Processing Technology* 2001, *71* (1), 139–148. https://doi.org/10.1016/S0378-3820(01)00140-0

(6) Peña, M. A.; Gómez, J. P.; Fierro, J. L. G. New Catalytic Routes for Syngas and Hydrogen Production. *Applied Catalysis A: General* 1996, *144* (1), 7–57. https://doi.org/10.1016/0926-860X(96)00108-1

(7) Review of Syngas Production via Biomass DFBGs. *Renewable and Sustainable Energy Reviews* 2011, *15* (1), 482–492. https://doi.org/10.1016/j.rser.2010.09.032

(8) Zhao, L.; Etezadi, R.; Tsotsis, T. Chapter 15 - Syngas from Coal. In *Advances in Synthesis Gas : Methods, Technologies and Applications*; Rahimpour, M. R., Makarem, M. A., Meshksar, M., Eds.; Elsevier, 2023; Vol. 1, pp 363–377. https://doi.org/10.1016/B978-0-323-91871-8.00013-1

(9) Chorkendorff, I.; Niemantsverdriet, J. W. *Concepts of Modern Catalysis and Kinetics*; John Wiley & Sons, 2017.

(10) Arunajatesan, V.; Wilson, K. A.; Subramaniam, B. Pressure-Tuning the Effective Diffusivity of Near-Critical Reaction Mixtures in Mesoporous Catalysts. *Ind. Eng. Chem. Res.* 2003, *42* (12), 2639–2643. https://doi.org/10.1021/ie020835d

(11) Auerbach, S. M.; Carrado, K. A.; Dutta, P. K. *Handbook of Zeolite Science and Technology*; CRC Press, 2003.

(12) Hagen, J. *Industrial Catalysis: A Practical Approach*; John Wiley & Sons, 2015.

(13) Kang, J.; Cheng, K.; Zhang, L.; Zhang, Q.; Ding, J.; Hua, W.; Lou, Y.; Zhai, Q.; Wang, Y. Mesoporous Zeolite-Supported Ruthenium Nanoparticles as Highly Selective Fischer–Tropsch Catalysts for the Production of C5–C11 Isoparaffins. *Angewandte Chemie International Edition* 2011, *50* (22), 5200–5203. https://doi.org/10.1002/anie.201101095

(14) Cheng, K.; Kang, J.; Huang, S.; You, Z.; Zhang, Q.; Ding, J.; Hua, W.; Lou, Y.; Deng, W.; Wang, Y. Mesoporous Beta Zeolite-Supported Ruthenium Nanoparticles for Selective Conversion of Synthesis Gas to C5–C11 Isoparaffins. *ACS Catal.* 2012, *2* (3), 441–449. https://doi.org/10.1021/cs200670j

(15) Sun, J.; Li, X.; Taguchi, A.; Abe, T.; Niu, W.; Lu, P.; Yoneyama, Y.; Tsubaki, N. Highly-Dispersed Metallic Ru Nanoparticles Sputtered on H-Beta Zeolite for Directly Converting Syngas to Middle Isoparaffins. *ACS Catal.* 2014, *4* (1), 1–8. https://doi.org/10.1021/cs4008842

(16) Stranges, A. Germany's Synthetic Fuel Industry, 1927–1945. In *The German Chemical Industry in the Twentieth Century*; Lesch, J. E., Ed.; Springer Netherlands: Dordrecht, 2000; pp 147–216. https://doi.org/10.1007/978-94-015-9377-9_7

(17) Vogel, A. P.; van Dyk, B.; Saib, A. M. GTL Using Efficient Cobalt Fischer-Tropsch Catalysts. *Catalysis Today* 2016, *259*, 323–330. https://doi.org/10.1016/j.cattod.2015.06.018

(18) Henrici-Olivé, G.; Olivé, S. The Fischer-Tropsch Synthesis: Molecular Weight Distribution of Primary Products and Reaction Mechanism. *Angewandte Chemie International Edition in English* 1976, *15* (3), 136–141. https://doi.org/10.1002/anie.197601361

(19) Dry, M. E. Catalytic Aspects of Industrial Fischer-Tropsch Synthesis. *Journal of Molecular Catalysis* 1982, *17* (2), 133–144. https://doi.org/10.1016/0304-5102(82)85025-6

(20) Jiao, F.; Li, J.; Pan, X.; Xiao, J.; Li, H.; Ma, H.; Wei, M.; Pan, Y.; Zhou, Z.; Li, M.; Miao, S.; Li, J.; Zhu, Y.; Xiao, D.; He, T.; Yang, J.; Qi, F.; Fu, Q.; Bao, X. Selective Conversion of Syngas to Light Olefins. *Science* 2016, *351* (6277), 1065–1068. https://doi.org/10.1126/science.aaf1835

(21) Cheng, K.; Zhou, W.; Kang, J.; He, S.; Shi, S.; Zhang, Q.; Pan, Y.; Wen, W.; Wang, Y. Bifunctional Catalysts for One-Step Conversion of Syngas into Aromatics with Excellent Selectivity and Stability. *Chem* 2017, *3* (2), 334–347. https://doi.org/10.1016/j.chempr.2017.05.007

(22) Xu, Y.; Ma, G.; Bai, J.; Du, Y.; Qin, C.; Ding, M. Yolk@Shell FeMn@Hollow HZSM-5 Nanoreactor for Directly Converting Syngas to Aromatics. *ACS Catal.* 2021, *11* (8), 4476–4485. https://doi.org/10.1021/acscatal.0c05658

(23) Liu, C.; Su, J.; Liu, S.; Zhou, H.; Yuan, X.; Ye, Y.; Wang, Y.; Jiao, W.; Zhang, L.; Lu, Y.; Wang, Y.; He, H.; Xie, Z. Insights into the Key Factor of Zeolite Morphology on the Selective Conversion of Syngas to Light Aromatics over a Cr_2O_3/ZSM-5 Catalyst. *ACS Catal.* 2020, *10* (24), 15227–15237. https://doi.org/10.1021/acscatal.0c03658

(24) Sartipi, S.; Alberts, M.; Meijerink, M. J.; Keller, T. C.; Pérez-Ramírez, J.; Gascon, J.; Kapteijn, F. Towards Liquid Fuels from Biosyngas: Effect of Zeolite Structure in Hierarchical-Zeolite-Supported Cobalt Catalysts. *ChemSusChem* 2013, *6* (9), 1646–1650. https://doi.org/10.1002/cssc.201300339

(25) Sineva, L. V.; Asalieva, E. Yu.; Mordkovich, V. Z. Role of Zeolite in the Synthesis of Liquid Hydrocarbons from CO and H_2 on a Composite Cobalt Catalyst. *Catal. Ind.* 2015, *7* (4), 245–252. https://doi.org/10.1134/S2070050415040145

(26) Wang, S.; Li, X.; Zhang, F.; Cai, Q.; Wang, Y.; Luo, Z. Bio-Oil Catalytic Reforming without Steam Addition: Application to Hydrogen Production and Studies on Its Mechanism. *International Journal of Hydrogen Energy* 2013, *38* (36), 16038–16047. https://doi.org/10.1016/j.ijhydene.2013.10.032

(27) Wang, S.; Cai, Q.; Wang, X.; Zhang, L.; Wang, Y.; Luo, Z. Biogasoline Production from the Co-Cracking of the Distilled Fraction of Bio-Oil and Ethanol. *Energy Fuels* 2014, *28* (1), 115–122. https://doi.org/10.1021/ef4012615

(28) Sartipi, S.; van Dijk, J. E.; Gascon, J.; Kapteijn, F. Toward Bifunctional Catalysts for the Direct Conversion of Syngas to Gasoline Range Hydrocarbons: H-ZSM-5 Coated Co *versus* H-ZSM-5 Supported Co. *Applied Catalysis A: General* 2013, *456*, 11–22. https://doi.org/10.1016/j.apcata.2013.02.012

(29) Sartipi, S.; Alberts, M.; Santos, V. P.; Nasalevich, M.; Gascon, J.; Kapteijn, F. Insights into the Catalytic Performance of Mesoporous H-ZSM-5-Supported Cobalt in Fischer–Tropsch Synthesis. *ChemCatChem* 2014, *6* (1), 142–151. https://doi.org/10.1002/cctc.201300635

(30) Sartipi, S.; Parashar, K.; Valero-Romero, M. J.; Santos, V. P.; van der Linden, B.; Makkee, M.; Kapteijn, F.; Gascon, J. Hierarchical H-ZSM-5-Supported Cobalt for the Direct Synthesis of Gasoline-Range Hydrocarbons from Syngas: Advantages, Limitations, and Mechanistic Insight. *Journal of Catalysis* 2013, *305*, 179–190. https://doi.org/10.1016/j.jcat.2013.05.012

(31) Zhuo, Y.; Zhu, L.; Liang, J.; Wang, S. Selective Fischer-Tropsch Synthesis for Gasoline Production over Y, Ce, or La-Modified Co/H-β. *Fuel* 2020, *262*, 116490. https://doi.org/10.1016/j.fuel.2019.116490

(32) Zhu, C.; Bollas, G. M. Gasoline Selective Fischer-Tropsch Synthesis in Structured Bifunctional Catalysts. *Applied Catalysis B: Environmental* 2018, *235*, 92–102. https://doi.org/10.1016/j.apcatb.2018.04.063

(33) Li, Z.; Wu, L.; Han, D.; Wu, J. Characterizations and Product Distribution of Co-Based Fischer-Tropsch Catalysts: A Comparison of the Incorporation Manner. *Fuel* 2018, *220*, 257–263. https://doi.org/10.1016/j.fuel.2018.02.004

(34) Sineva, L. V.; Gorokhova, E. O.; Gryaznov, K. O.; Ermolaev, I. S.; Mordkovich, V. Z. Zeolites as a Tool for Intensification of Mass Transfer on the Surface of a Cobalt Fischer–Tropsch Synthesis Catalyst. *Catalysis Today* 2021, *378*, 140–148. https://doi.org/10.1016/j.cattod.2021.02.018

(35) Li, X.; Chen, Y.; Liu, S.; Zhao, N.; Jiang, X.; Su, M.; Li, Z. Enhanced Gasoline Selectivity through Fischer-Tropsch Synthesis on a Bifunctional Catalyst: Effects of Active Sites Proximity and Reaction Temperature. *Chemical Engineering Journal* 2021, *416*, 129180. https://doi.org/10.1016/j.cej.2021.129180

(36) Li, N.; Jiao, F.; Pan, X.; Chen, Y.; Feng, J.; Li, G.; Bao, X. High-Quality Gasoline Directly from Syngas by Dual Metal Oxide–Zeolite (OX-ZEO) Catalysis. *Angewandte Chemie International Edition* 2019, *58* (22), 7400–7404. https://doi.org/10.1002/anie.201902990

(37) Wang, C.; Bu, X.; Ma, J.; Liu, C.; Chou, K.; Wang, X.; Li, Q. Wells–Dawson Type $Cs_{5.5}H0.5P_2W_{18}O_{62}$ Based Co/Al_2O_3 as Bifunctional Catalysts for Direct Production of Clean-Gasoline Fuel through Fischer–Tropsch Synthesis. *Catalysis Today* 2016, *274*, 82–87. https://doi.org/10.1016/j.cattod.2016.01.043

(38) Cheng, K.; Zhou, W.; Kang, J.; He, S.; Shi, S.; Zhang, Q.; Pan, Y.; Wen, W.; Wang, Y. Bifunctional Catalysts for One-Step Conversion of Syngas into Aromatics with Excellent Selectivity and Stability. *Chem* 2017, *3* (2), 334–347. https://doi.org/10.1016/j.chempr.2017.05.007

(39) Pastor-Pérez, L.; Buitrago-Sierra, R.; Sepúlveda-Escribano, A. CeO$_2$-Promoted Ni/Activated Carbon Catalysts for the Water–Gas Shift (WGS) Reaction. *International Journal of Hydrogen Energy* 2014, *39* (31), 17589–17599. https://doi.org/10.1016/j.ijhydene.2014.08.089

(40) Wang, Y.; Yu, J.; Qiao, J.; Sun, Y.; Jin, W.; Zhang, H.; Ma, J. Effect of Mesoporous ZSM-5 Morphology on the Catalytic Performance of Cobalt Catalyst for Fischer-Tropsch Synthesis. *Journal of the Energy Institute* 2020, *93* (3), 1187–1194. https://doi.org/10.1016/j.joei.2019.11.002

(41) Owen, N. A.; Inderwildi, O. R.; King, D. A. The Status of Conventional World Oil Reserves—Hype or Cause for Concern? *Energy Policy* 2010, *38* (8), 4743–4749. https://doi.org/10.1016/j.enpol.2010.02.026

(42) Liederman, D.; Yurchak, S.; Kuo, J. C. W.; Lee, W. Mobil Methanol-to-Gasoline Process. *Journal of Energy* 1982, *6* (5), 340–341. https://doi.org/10.2514/3.62614

(43) Olsbye, U.; Svelle, S.; Bjørgen, M.; Beato, P.; Janssens, T. V. W.; Joensen, F.; Bordiga, S.; Lillerud, K. P. Conversion of Methanol to Hydrocarbons: How Zeolite Cavity and Pore Size Controls Product Selectivity. *Angewandte Chemie* International Edition 2012, *51* (24), 5810–5831. https://doi.org/10.1002/anie.201103657

(44) Deka, T. J.; Osman, A. I.; Baruah, D. C.; Rooney, D. W. Methanol Fuel Production, Utilization, and Techno-Economy: A Review. *Environ Chem Lett* 2022, *20* (6), 3525–3554. https://doi.org/10.1007/s10311-022-01485-y

(45) Chang, C. D.; Kuo, J. C. W.; Lang, W. H.; Jacob, S. M.; Wise, J. J.; Silvestri, A. J. Process Studies on the Conversion of Methanol to Gasoline. *Ind. Eng. Chem. Proc. Des. Dev.* 1978, *17* (3), 255–260. https://doi.org/10.1021/i260067a008

(46) Kianfar, E.; Hajimirzaee, S.; Mousavian, S.; Mehr, A. S. Zeolite-Based Catalysts for Methanol to Gasoline Process: A Review. *Microchemical Journal* 2020, *156*, 104822. https://doi.org/10.1016/j.microc.2020.104822

(47) Haw, J. F.; Song, W.; Marcus, D. M.; Nicholas, J. B. The Mechanism of Methanol to Hydrocarbon Catalysis. *Acc. Chem. Res.* 2003, *36* (5), 317–326. https://doi.org/10.1021/ar020006o

(48) Hutchings, G. J.; Hunter, R. Hydrocarbon Formation from Methanol and Dimethyl Ether: A Review of the Experimental Observations Concerning the Mechanism of Formation of the Primary Products. *Catalysis Today* 1990, *6* (3), 279–306. https://doi.org/10.1016/0920-5861(90)85006-A

(49) Stöcker, M. Methanol-to-Hydrocarbons: Catalytic Materials and Their Behavior1. *Microporous and Mesoporous Materials* 1999, *29* (1), 3–48. https://doi.org/10.1016/S1387-1811(98)00319-9

(50) Tabak, S. A.; Yurchak, S. Conversion of Methanol over ZSM-5 to Fuels and Chemicals. *Catalysis Today* 1990, *6* (3), 307–327. https://doi.org/10.1016/0920-5861(90)85007-B

(51) Zaidi, H. A.; Pant, K. K. Catalytic Conversion of Methanol to Gasoline Range Hydrocarbons. *Catalysis Today* 2004, *96* (3), 155–160. https://doi.org/10.1016/j.cattod.2004.06.123

(52) Yurchak, S. Development of Mobil's Fixed-Bed Methanul-to-Gasoline (MTG) Process. In *Studies in Surface Science and Catalysis*; Bibby, D. M., Chang, C. D., Howe, R. F., Yurchak, S., Eds.; Methane Conversion; Elsevier, 1988; Vol. 36, pp 251–272. https://doi.org/10.1016/S0167-2991(09)60521-8

(53) Chang, C. D. The New Zealand Gas-to-Gasoline Plant: An Engineering Tour de Force. *Catalysis Today* 1992, *13* (1), 103–111. https://doi.org/10.1016/0920-5861(92)80190-X

(54) Vora, B. V.; Marker, T. L.; Barger, P. T.; Nilsen, H. R.; Kvisle, S.; Fuglerud, T. Economic Route for Natural Gas Conversion to Ethylene and Propylene. In *Studies in Surface Science and Catalysis*; de Pontes, M., Espinoza, R. L., Nicolaides, C. P., Scholtz, J. H., Scurrell, M. S., Eds.; Natural Gas Conversion IV; Elsevier, 1997; Vol. 107, pp 87–98. https://doi.org/10.1016/S0167-2991(97)80321-7

(55) Koempel, H.; Liebner, W. Lurgi's Methanol To Propylene (MTP®) Report on a Successful Commercialisation. In *Studies in Surface Science and Catalysis*; Bellot Noronha, F., Schmal, M., Falabella Sousa-Aguiar, E., Eds.; Natural Gas Conversion VIII; Elsevier, 2007; Vol. 167, pp 261–267. https://doi.org/10.1016/S0167-2991(07)80142-X

(56) Kianfar, E. Ethylene to Propylene over Zeolite ZSM-5: Improved Catalyst Performance by Treatment with CuO. *Russ J Appl Chem* 2019, *92* (7), 933–939. https://doi.org/10.1134/S1070427219070085

(57) Kianfar, E. Ethylene to Propylene Conversion over Ni-W/ZSM-5 Catalyst. *Russ J Appl Chem* 2019, *92* (8), 1094–1101. https://doi.org/10.1134/S1070427219080068

(58) Wilson, S.; Barger, P. The Characteristics of SAPO-34 Which Influence the Conversion of Methanol to Light Olefins. *Microporous and Mesoporous Materials* 1999, *29* (1), 117–126. https://doi.org/10.1016/S1387-1811(98)00325-4

(59) Höchtl, M.; Jentys, A.; Vinek, H. Isomerization of 1-Pentene over SAPO, CoAPO (AEL, AFI) Molecular Sieves and HZSM-5. *Applied Catalysis A: General* 2001, *207* (1), 397–405. https://doi.org/10.1016/S0926-860X(00)00682-7

(60) Sun, X.; Mueller, S.; Liu, Y.; Shi, H.; Haller, G. L.; Sanchez-Sanchez, M.; van Veen, A. C.; Lercher, J. A. On Reaction Pathways in the Conversion of Methanol to Hydrocarbons on HZSM-5. *Journal of Catalysis* 2014, *317*, 185–197. https://doi.org/10.1016/j.jcat.2014.06.017

(61) Kianfar, E. A Comparison and Assessment on Performance of Zeolite Catalyst Based Selective for the Process Methanol to Gasoline: *A Review*; 2020.

(62) Su, X.; Liu, B.; Feng, C.; Wu, W. Rapid and Low-Cost Synthesis of ZSM-5 Zeolite Nanosheet Assemblies for Conversion of Methanol to Gasoline. *Microporous and Mesoporous Materials* 2022, *344*, 112215. https://doi.org/10.1016/j.micromeso.2022.112215

(63) Jia, Y.; Wang, J.; Zhang, K.; Feng, W.; Liu, S.; Ding, C.; Liu, P. Nanocrystallite Self-Assembled Hierarchical ZSM-5 Zeolite Microsphere for Methanol to Aromatics. *Microporous and Mesoporous Materials* 2017, *247*, 103–115. https://doi.org/10.1016/j.micromeso.2017.03.035

(64) Lyu, Y.; Zhan, W.; Yu, Z.; Liu, X.; Yang, Y.; Wang, X.; Song, C.; Yan, Z. One-Pot Synthesis of the Highly Efficient Bifunctional Ni-SAPO-11 Catalyst. *Journal of Materials Science & Technology* 2021, *76*, 86–94. https://doi.org/10.1016/j.jmst.2020.10.033

(65) Wan, Z.; Wu, W.; Li, G. (Kevin); Wang, C.; Yang, H.; Zhang, D. Effect of SiO_2/Al_2O_3 Ratio on the Performance of Nanocrystal ZSM-5 Zeolite Catalysts in Methanol to Gasoline Conversion. *Applied Catalysis A: General* 2016, *523*, 312–320. https://doi.org/10.1016/j.apcata.2016.05.032

(66) Mirshafiee, F.; Karimzadeh, R.; Khoshbin, R. Free Template Synthesis of Novel Hybrid MFI/BEA Zeolite Structure Used in the Conversion of Methanol to Clean Gasoline: Effect of Beta Zeolite Content. *Fuel* 2021, *304*, 121386. https://doi.org/10.1016/j.fuel.2021.121386

(67) Kianfar, E.; Salimi, M.; Koohestani, B. Methanol to Gasoline Conversion over CuO/ZSM-5 Catalyst Synthesized and Influence of Water on Conversion: *Fine Chemical Engineering* 2020, 75–82. https://doi.org/10.37256/fce.122020499

(68) Di, Z.; Yang, C.; Jiao, X.; Li, J.; Wu, J.; Zhang, D. A ZSM-5/MCM-48 Based Catalyst for Methanol to Gasoline Conversion. *Fuel* 2013, *104*, 878–881. https://doi.org/10.1016/j.fuel.2012.09.079

(69) Meng, F.; Wang, Y.; Wang, X. Methanol to Gasoline over ZSM-5 Zeolite Treated with Alkaline Mixture with Different Ratios of TBA+/OH−. *Kinet Catal* 2018, *59* (3), 304–310. https://doi.org/10.1134/S0023158418030138

(70) Li, J.; Miao, P.; Li, Z.; He, T.; Han, D.; Wu, J.; Wang, Z.; Wu, J. Hydrothermal Synthesis of Nanocrystalline H[Fe, Al]ZSM-5 Zeolites for Conversion of Methanol to Gasoline. *Energy Conversion and Management* 2015, *93*, 259–266. https://doi.org/10.1016/j.enconman.2015.01.031

(71) Meng, F.; Wang, Y.; Wang, S.; Wang, X.; Wang, S. Synthesis of ZSM-5 Aggregates by a Seed-Induced Method and Catalytic Performance in Methanol-to-Gasoline Conversion. *Comptes Rendus Chimie* 2017, *20* (4), 385–394. https://doi.org/10.1016/j.crci.2016.07.005

(72) Sanz-Martínez, A.; Lasobras, J.; Soler, J.; Herguido, J.; Menéndez, M. Methanol to Gasoline (MTG): Preparation, Characterization and Testing of HZSM-5 Zeolite-Based Catalysts to Be Used in a Fluidized Bed Reactor. *Catalysts* 2022, *12* (2), 134. https://doi.org/10.3390/catal12020134

(73) Sun, Y.; Han, S. Mechanistic Investigation of Methanol to Propene Conversion Catalyzed by H-Beta Zeolite: A Two-Layer ONIOM Study. *J Mol Model* 2013, *19* (12), 5407–5422. https://doi.org/10.1007/s00894-013-2030-6

(74) Choe, J.; Choe, C.; Kim, T.; Pak, Y.; Han, C.; Yun, H. Novel Kinetic Modelling of Methanol-to-Gasoline (MTG) Reaction on HZSM-5 Catalyst: Product Distribution. *Journal of the Indian Chemical Society* 2021, *98* (2), 100003. https://doi.org/10.1016/j.jics.2021.100003

10 The Future of the Production of Fuels

Will Gasoline Survive the Electric Fuels?

Natalya Victorovna Likhanova
and Silvia Castillo-Acosta
Mexican Petroleum Institute, Mexico City, México

10.1 INTRODUCTION

In 2023 the European Union (EU) made the decision of reducing by 2035 automobile carbon dioxide emissions to practically zero as part of a program to lower the emissions of greenhouse gases (GHGs) by at least 55%. The achievement of this goal would mean that cars powered by fossil gasoline and diesel would disappear within 12 years whereas electric cars would become the prevailing vehicles up to 99%. However, currently, out of 249 million cars in the 27 countries that are part of the EU, only 0.5% correspond to totally electric vehicles. As for world traffic, the participation of electric cars is not higher than 3% in countries leading in the use of electric cars: China, the United States, Norway, and Germany (by 2030).[1] In this EU treaty, where the use of new vehicles with internal combustion engines will be banned from 2035, the German Federal Minister of Transport and Digital Infrastructure, Volker Wissing, added the following exception: after 2035, automobiles with diesel and gasoline engines will be allowed to use only fuel types bearing the labels "environmentally neutral" or "electrofuel" or "e-fuel" in order to promote a smooth temporal transition toward electric vehicles.[2,3]

E-fuels are an emerging class of carbon-neutral synthetic fuels that are produced by employing either captured carbon dioxide or nitrogen, along with hydrogen produced from water electrolysis or photolysis powered by renewable electricity. Sustainable energy is based on the production and storage of energy with negligible collateral effects, mainly on the environment. In this context, even the African continent has joined the "Green Agenda" and the production of "clean" energy. African countries like Kenya, Ethiopia, Namibia, Rwanda, Sierra Leone, and Zimbabwe with the help of Denmark, Germany, and the United Arab Emirates founded a new partnership called "Accelerated Partnership for Renewables in Africa," which in September 2023 announced the goal to achieve 100% of renewable power by 2030 and to fuel the green industries of the future by 2040.[4] In general, "clean" energy is classified according to production methods such as solar, wind, geothermal, and hydro. Each sustainable energy production method has its pros and cons and depends on the region where it is produced. For instance, the world's leading producer of geothermal energy by 2023 will be the United States, followed by Indonesia and the Philippines, while Mexico is in the seventh place,[5,6] whereas the main producers of hydroelectric energy are China, Brazil, Canada, the United States, and Russia because of the presence of mighty rivers in their territory, although they have to be dammed, which alters completely the environment.[7] Wind energy is produced mainly in China, the United Kingdom, Germany, the Netherlands, and Denmark.[8-10] The use of biomass as a fuel source depends directly on every specific region, where many successful cases can be found like the one of bioethanol in Brazil, where up to 11% of vehicles use this type of fuel.[11,12] In this context, it is worth mentioning that bioethanol as fuel for transport engines is a quite old idea. During the First World War, combat planes employed ethanol, obtained from potatoes, as fuel. In Europe, until the 1950s, bioethanol was mixed with existing fuels such as gasoline, benzene, methanol, acetone, etc. to improve the antiknock properties. Up to now, in the EU, from 10% to 85% of ethanol is added to gasoline to emit less carbon dioxide.[13] Since 2005, the interest in using bioethanol and the technology of internal combustion engines powered by ethanol has increased. Notwithstanding, sometimes, the production of biofuels requires huge amounts of water and fertilizers, which leads to deforestation and biodiversity loss in the production zone.[14]

The case of atomic energy is very interesting, for nuclear plants do not produce GHGs. An one-inch-enriched-uranium pellet produces energy equivalent to 1 ton of carbon or 120 oil barrels. In addition, last year, Russia was the first country in the world to accomplish a closed nuclear cycle, which implies the recycling and reuse of nuclear fuel. Fast-neutron reactors can prepare new fuel for themselves and if necessary, for thermal reactors, as well as burn uranium from spent nuclear fuel that remains in the thermal reactors. This scheme is called "two-component nuclear energy," where the fast-neutron and thermal reactors work together by exchanging fuel. Then, the resource base for the development of nuclear energy becomes almost unlimited.[15] The very design of the innovative fast-neutron reactor BREST-OD-300 developed by the state Russian company Rosatom excludes the occurrence of hypothetical accidents in the nuclear plants; from the economic point of view, closed-cycle energy systems will be capable of fully exploiting the energy potential of natural uranium up to 99% instead of the current 0.7% by forming neptunium and curium.[16] Based on the foregoing, nuclear energy can be classified as sustainable energy.

Finally, e-fuels can be green hydrogen, ammonia, or synthesis products from carbon dioxide and hydrogen such as methane, methanol, their derivatives, or mixtures of "liquid" alkanes and olefins (Figure 10.1).

It is true that by refueling automobiles with synthetic fuels, emissions of GHGs are not prevented, but they are much lower than when fossil-origin gasoline and diesel are employed. Then, according to the Union of German Motorists (ADAC), the gas emission (nitrous oxide and carbon dioxide) level diminished by 39% when synthetic fuels were used.[17] This fact will allow to continue the operation of internal combustion engines, which will be very important for millions of owners of this type of car.

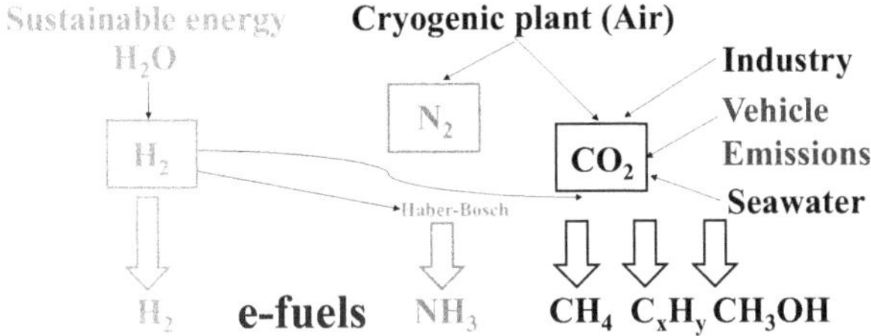

FIGURE 10.1 E-fuel types and their sources.

An ideal environmental scenario would be to employ "green energy" directly in the cars without having to pass through the internal combustion engine and in this way avoid using any liquid or gaseous fuel; for this reason, the presence of photovoltaic cells in automobiles could be a feasible and very "green" option.

10.2 SOLAR ENERGY

The regions with optimal ecological conditions and intense solar radiation for the local production of electrofuels are located at the 35th parallel north and south, where solar electricity can be produced with efficiency rates by electricity production facilities that are three or four times higher than those achieved in Central Europe or North America, mainly throughout the Atacama Desert plateau. Atacama is a part of countries such as Argentina, Bolivia, Chile, and Peru and is the place on the planet that receives higher surface solar irradiation.[18,19] In 2023, several big projects involving photovoltaic parks were carried out in the world. Thus, in November 2023, in the United Arab Emirates, one of the biggest solar parks in the world, Al Dhafra, which is located about 35 km from Abu Dhabi, started operations. The capacity of this electric power plant is 2 GW, which is enough to satisfy the energy demand of almost 200,000 homes on average.[20,21] Likewise, in another part of the world, also in November, in Indonesia, at 108 km from the capital Jakarta, the biggest floating solar energy plant in Southeast Asia with an approximate capacity of 1 GW was inaugurated to power 50,000 homes per year.[22–24] The Indian company Adani Green Energy Limited announced the completion of a solar–wind energy plant with a capacity of 2.2 GW for powering 100,000 homes.[25] In Latin America, in the electricity production branch by solar energy, Chile has the first place; however, in absolute numbers, by 2022, Brazil won the first place with 832 GWh, followed by Mexico with 5.52 kWh/kWp.[26] Mexico has 39 operational solar parks, where two-thirds of the energy capacity is located in four states: Baja California Sur, Durango, Chihuahua, and the State of Mexico. Currently, in Puerto Peñasco, in Sonora desert, the biggest photovoltaic power plant in Latin America with 2,000 hectares of solar panels is being constructed and by 2027, it will produce electricity equivalent to 1 GW.[27,28] Even Mexico City has the biggest photovoltaic solar plant in the world installed on a roof, which produces energy for the Supply Center (Central de Abastos) in the Iztapalapa Mayoralty.[29]

The problem faced by these systems is that there is a big loss of electric energy produced by the long distances that have to be covered until reaching the consumers.

The use of solar cells to produce electricity directly for automobiles avoids the winding path between the production point and energy demand point, which is being searched for by the automobile industry.[30] In this context, in 2008, the company Fisker Automotive released to the market the Fisker Karma, the first plug-in electric car in the world with large solar roof panels.[31] However, the efficiency of the solar cells was still low and automobiles with solar cells could produce enough energy for up to 8 km of traveled distance. By the end of December 2022, Valmet Automotive and the Dutch company Lightyear, after six years of research and development, started the massive production of the first automobile in the world that received its energy from a solar-panel-based system. Each detail of the vehicle was carefully thought out to minimize energy consumption and maximize its autonomy. The car has a record drag coefficient and the surface is covered with 5 m^2 of double-curved solar panels. With 170 hp, this car, which costs more than 258 550 USD, on a sunny day, is capable of generating, only thanks to the sun, enough energy to travel 70 km.[32,33] In this context, the company Aptera launched an electric vehicle with solar panels, which consumes 30% of energy in comparison to other electric and hybrid vehicles.[34] Furthermore, there are many prototypes of electric cars with solar panels at the research level that are being developed mainly by researchers from countries where there is intense solar radiation like Mexico, Greece, Pakistan, and India.[35–37]

To understand why the efficiency of solar cells, up to now, is not higher than 33%, we will see what they are made of. Currently research works related to solar cells look for materials with band gap values between 1.1 and 1.7 eV (the silicon band gap is around 1.12 eV), for they facilitate the promotion of an electron from one band to the other, thus increasing material conduction.[38] There is a great variety of photovoltaic cell technologies, which can be classified into three generations depending on the employed raw materials and the level of commercial maturity.[39–42] Figure 10.2 summarizes this information as follows:

1. Silicon-based photovoltaic cells. These cells are classified as monocrystalline (m-Si), polycrystalline (pc-Si), and amorphous. Monocrystalline systems are among the most popular and their efficiency varies from 18% to 26% and average useful life can reach 25 years,[43] although the production cost is high. During the manufacturing process, a special crystal is placed in molten silicon and then cut into fine rectangles. A negative aspect is performance reduction under low illumination conditions due to the peculiarity of the directionality of silicon particles. Polycrystalline cells are made by pouring molten silicon into molds. The efficiency of polycrystalline cells varies from 15% to 20%,[44,45] but the cost is between 10% and 20% lower than that of monocrystalline models.[46] Due to the random disposition of silicon microparticles, they are capable of generating energy under low illumination or cloudy conditions. At present, amorphous solar cells are hardly used due to their low efficiency.
2. Solar cells with a semiconductor component are thin-film panels made of tellurium, cadmium, indium, gallium, and copper selenide materials. These photovoltaic cells require much less semiconductor material[47] and are highly flexible and easily installed; however, their efficiency is not higher

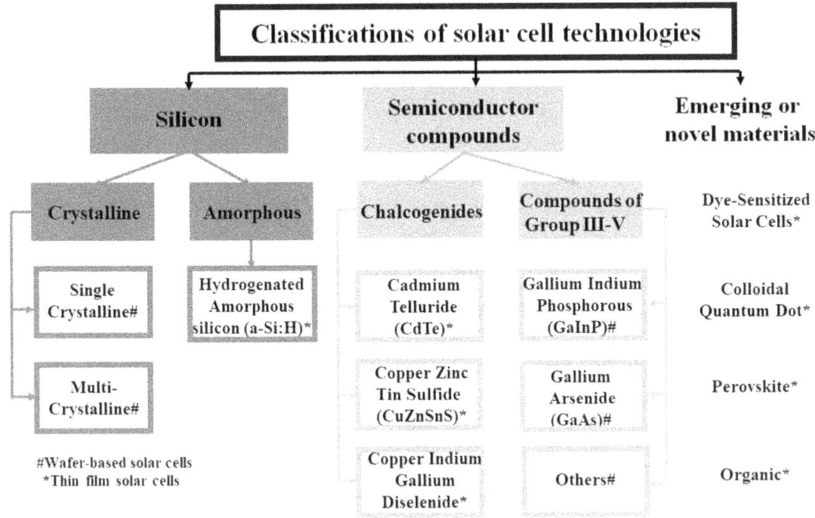

FIGURE 10.2 Types of solar cells. Reprinted from The Lancet, [65] Vol. 80, T. Ibn-Mohammed, S.C.L. Koh, I.M. Reaney, A. Acquaye, G. Schileo, K.B. Mustapha, R. Greenough, Perovskite solar cells: An integrated hybrid lifecycle assessment and review in comparison with other photovoltaic technologies, *Renewable and Sustainable Energy Reviews*, Pages 1321–1344, Copyright (2017), with permission from Elsevier.

than 22%.[48–50] Nowadays, gallium-arsenide-based cells are being used and the direct bandgap of GaAs is 1.42 eV at 27°C; these cells are considerably expensive and an approximate efficiency of 30% has been recorded.[51–54]

3. "Emerging photovoltaic cells." This group is represented by organic nano-structured and mesostructured photovoltaic cells (e.g., dye-sensitized[55,56] and quantum dot solar cells[57–59]) based on polymeric semiconductors (poly-phenylene vinylene and copper phthalocyanine) and carbon-based nano-structures (fullerenes, nanotubes, and graphene), which have a band gap value above 2.0 eV that limits the efficient absorption of solar photons.[60] Due to the fact that their efficiency is normally below 13%, these cells were never produced industrially.[61,62] Currently, the scientific attention is focused on perovskite- and organometallic-halide-based solar cells, which have displayed efficiency as high as 26%, high structural stability, favorable optoelectrical properties, and easy production features.[63–65]

Even if the efficiency issue had been solved with semiconductor materials with efficiency up to 35%, the inclusion of solar cells in electric cars would have been beneficial for short distances but not for long ones.[35]

Although modern electric automobiles feature several advantages such as no emission of exhaust gases and relatively easy fixing, they also have several disadvantages like limited autonomy – normally, an electric car cannot be at distances farther than 135 km from the charge port, the charging process is extremely long (between 4 and 10 h, approximately), and they have power and speed limitations; in cold weather, electric cars lose two-thirds of their efficiency; currently, the price/power ratio is very distorted and finally, there are problems with lithium batteries.[66,67] Although the price of batteries has been reduced by 85% from 2010,[68] lithium has worse limitations than oil for it is a scarce resource and is mainly found in South America, in countries such as Argentina, Bolivia, Chile, and Peru, especially in Bolivia.[69,70] A car battery emits around 7.5 tons of carbon dioxide when produced.[71] The average useful life of a lithium battery is 3000 charge cycles and loses its capacity very quickly; for example, in a 100-kW battery, The loss of around 20% of its capacity is not a grave problem, but in a 40-kWh battery, which is normally used in small electric automobiles, the performance loss would be of 32 kWh, which would be very important. Furthermore, the current recycling process is highly pollutant and complicated. For this reason, first, lithium batteries should be completely discharged in order to be put to pieces and separate their components; spent batteries are destined for other uses such as illuminating football fields, streets, etc., where neither the fast charge and discharge nor the energy density is important. Afterward, the discharged batteries are ground and acids and high temperatures are employed for recovering metals, although just 20–80% of components are recovered. This procedure is very expensive and generates considerable CO_2 emissions.[71–73]

Due to the panorama described above, the transition to electric vehicles, mainly in the EU, can only be possible by using electrofuel.

10.3 E-FUEL BASED ON CO_2

Examples of e-fuels from CO_2 not only include methane, methanol, and their derivatives (mainly dimethyl ether) but also more complex hydrocarbons like gasoline and kerosene. It is worth mentioning that when using methane as e-fuel, it is necessary to follow extreme security precautions to prevent leaks, for methane is almost 26.5 times more powerful than CO_2 in terms of global warming.[74] Synthetic hydrocarbons are produced at temperatures between 200 and 300°C and at pressures between 20 and 50 bar, using catalysts and an average of 3 moles of hydrogen per 1 mole of carbon dioxide, producing water as a subproduct. The chemical reactions and details concerning the production of e-fuels from carbon dioxide were reviewed in previous chapters of this book.[75–77]

The cheapest source of carbon dioxide for this application is flue gas, which comes from burning fossil fuels and can be obtained for 7.50 USD a ton. Strictly speaking, this is not a carbon-neutral process, for carbon has fossil origin. The capture of vehicle exhaust gases is also considered as profitable but requires design changes in automobiles. Since carbon dioxide in seawater is in chemical equilibrium with atmospheric carbon dioxide, carbon extraction from seawater is being studied. Researchers estimate that this extraction process will cost around 50 dollars a ton. Carbon capture from atmospheric air is more expensive, for it oscillates between 94 and 232 dollars a ton.[78,79] Finally, there is the possibility of employing biogas as a carbon source, obtained from anaerobic digestion. In addition, for more efficient use of biomass, hydrogen is added, which leads to the production of fuel.

Out of 19.3 million tons of methanol produced in 2022, 80% came from natural gas. Green methanol has an estimated cost of 680 dollars a ton and cannot compete against the price of fossil-origin methanol, which normally costs around 300 dollars a ton. The George Olah plant in Iceland became the first commercial facility capable of producing "green" methanol with low carbon emissions (known as "Vulcanol"). The plant produces 4000 tons of methanol per year. CO_2 is captured in a nearby geothermal energy plant, using low-cost electricity based on geothermal energy.[80]

Porsche, BMW, McLaren, and other companies started working on synthetic fuel in an effort to rescue gasoline engines. In December 2022, the Porsche partners – Siemens Energy, ExxonMobil AME, ENAP, and ENEL – opened a plant to produce e-fuel in Chile as part of the project "Haru Oni." This plant can produce 130,000 liters of e-fuel per year. The intensified operation will allow the production of 550,000 liters in 2027. To this end, the construction of at least 380 wind farms will be necessary.[81–83] Thus, the Norwegian company Norsk e-fuel started the production of 10,000 tons per year of synthetic kerosene using CO_2 extracted from the air, the technology of the Swiss company Climeworks, and hydroenergy from the fjords.[84,85]

10.4 HYDROGEN AS E-FUEL

The future of electric mobility lies in hydrogen fuel batteries and not in lithium ones. Although hydrogen is the most abundant element in the universe and any nation can have access to it, it is normally associated with other elements. For this reason, the technology and energy for separating molecular hydrogen is still highly costly.[86] The cost of obtaining hydrogen from renewable electricity, produced from water electrolysis or photolysis, is between 3.0 and 7.5 USD/Kg, which is considerably higher than that of natural gas that is around 1–2 USD/Kg or than that of carbon that is between 2 and 3 USD/Kg. Just after 2050, it is expected that the price of hydrogen could approach or equal that of natural gas.[87] The activity around hydrogen production is growing, for it is expected to obtain 30 million tons of low-emission hydrogen through the electrolysis technology around the world by 2030. In this context, China is planning to increase its participation in the production of low-emission hydrogen from 30% (currently) to 50% by 2024.[88] Photocatalysis is another viable alternative, but due to its low hydrogen production, much research work is being carried out to increase it, especially by employing visible light.[89] The global demand for hydrogen in 2023 was 53 million tons, it is expected that by 2050 it will be 500 to 680 million tons.[88] Hydrogen can be produced from renewable and nonrenewable sources like the following: natural gas, carbon, biomass, waste matter, solar sources, wind sources, and nuclear sources.

The hydrogen source determines if the energy system in question has no emissions in a clean network with a mitigated carbon footprint or if it pollutes by generating GHGs.[90]

Hydrogen can be classified according to the source and CO_2 emissions during its production:[91]

- Renewable hydrogen or green hydrogen: water is the hydrogen source, produced from renewable electricity for the electrolysis process. Another source is the reforming of biogas or biomass biochemical conversion.

- Blue hydrogen: natural gas, light hydrocarbons, or liquefied petroleum gases submitted to a reforming process are the hydrogen source; capture techniques and use and storage of carbon are employed (CCUS: Carbon Capture, Utilization, and Storage) in order to mitigate 95% of CO_2 emissions generated during the process.
- Grey hydrogen: natural gas, light hydrocarbons, or liquefied petroleum gases are the hydrogen source after a reforming process.
- Yellow hydrogen: water is the hydrogen source and is produced from the primary electricity network for the electrolysis process.
- Pink hydrogen: water is the hydrogen source and is produced from nuclear energy electricity for the electrolysis process.
- Turquoise hydrogen: pyrolysis-produced methane is the hydrogen source and does not require a carbon capture process, for solid carbon is generated.
- Black hydrogen: carbon is the hydrogen source, which has a high impact on the environment.
- Brown hydrogen: carbon is the hydrogen source.
- Gold hydrogen: biomethane reformed with water vapor, from which CO_2 is captured, is the hydrogen source.

Hydrogen storage is carried out in gaseous, liquid, and even solid forms in compressed gas tanks, cryogenic tanks, or caverns. Distribution is carried out by tank trucks or pipelines to the places where electric vehicles will be supplied to recharge the vehicles. The properties and safety characteristics of hydrogen make it necessary to adapt the current infrastructure for gas or make the infrastructure for use on par with fossil fuels. This infrastructure requires large investments in different segments of the industry and achieving profitability.[92] The use of hydrogen as e-fuel barely reached 0.03% in 2022.[88]

Considering that two-thirds of primary energy is used in transport and heating, the application of hydrogen to power transport represents important benefits due to its high efficiency and zero emissions.[93] The idea of using hydrogen as a fuel is not new. The hydrogen technology for zero-emission- and hybrid vehicles is a clean and sustainable energy system that can replace gasoline in the future in order to mitigate the emissions of GHGs and promote decarbonization;[94] in this way, the goals set in the Paris Agreement about the global temperature in 2015, which was adopted by 196 members participating at the Climate Change Conference (COP21), are on the way to be accomplished.[95] However, this technology is being developed by considering the production, storage, and distribution. Infrastructure with much higher investment than with electric models is required; in addition, the safety topic is very complicated, although the use of hybrid vehicles that combine internal combustion engines continues advancing by adapting itself to the existing infrastructure that is widely spread for gasoline internal combustion cars.[96] The key technology for the final use of hydrogen is represented by hydrogen fuel cells and it can be said that their commercialization is advancing quickly,[90] and this technology is already being employed in northern countries like Sweden.[97] As the cell technology is improved and made affordable, electric vehicles become more competitive than traditional ones with internal combustion engines in terms of research, autonomy, performance, and convenience.[95]

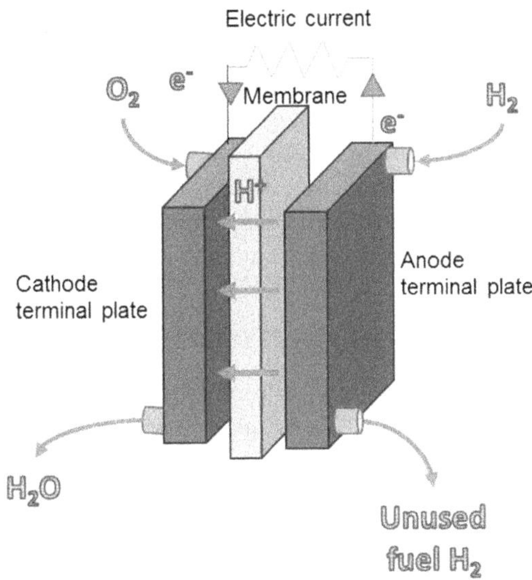

FIGURE 10.3 Hydrogen fuel cell.

The hydrogen supplied to an electric vehicle is kept in a storage cell that passes to a fuel cell where an electrochemical process of direct conversion of chemical energy into electrical energy is carried out through a chemical reaction between the hydrogen fuel source and oxygen. As shown in Figure 10.3, hydrogen is oxidized at the anode and the hydrogen protons generated are transported via the electrolyte medium to the cathode and the electrons through an external circuit are also transported to the cathode, where oxygen is fed and reacts with the hydrogen protons. The products of this reaction are water, electricity, and heat; therefore, there are zero emissions. The energy density of hydrogen is 143 MJ/kg, heat combustion is 141.9 MJ/kg, the calorific value is 120 MJ/kg, and the octane number is 130.[95]

The size of the hydrogen fuel tank determines the energy available to the electric vehicle.[98]

Automotive giants such as Audi, BMW, Nissan, Hyundai, Ford, Honda, and Toyota are producing limited editions of the Audi A7, BMW Series 7, and Mazda RX-8 that use the hydrogen fuel.[99,100] Probably the most famous model is the Toyota Mirai, which is a hybrid hydrogen car and is capable of loading 5 kg of hydrogen in 3 min for around 65.4 USD and consumes 1 kg of fuel per 100 km.[101]

The use of hydrogen as a fuel is part of the strategies set by governments for their energy transition; however, high demand comes from industrial applications such as oil refining and the steel industry;[88] on the other hand, many countries and regions are implementing stricter emission regulations and encouraging the acquisition of electric vehicles through subsidies, tax incentives, and infrastructure investment. These policies could accelerate the transition from gasoline-powered vehicles.[88] Examples are Japan and the United States with sea-floating wind projects that are the result of their partnership to reduce costs of profiting from high sea wind energy; this is an innovative approach to maximize efficiency. Japan invested 184 million dollars in their Green

Innovation Fund, thus promoting the implementation of efficient and strong floating turbines[102] for the production of hydrogen based on renewable-source electric energy.

The path leading to the future depends a lot on government incentives and the advances in the efficiency and technology of hydrogen fuel cells, supply models, production cost, packaging, and hydrogen transport. The hydrogen yearly cost in the hydrogen supply posts in California ranges from 13 to 15 US\$/kg of H_2. As the posts become more active and the demand increases, the leveled cost of the service station can be cut down to 2 US\$/kg of H_2 after a suitable scale economy.[93]

10.5 AMMONIA AS E-FUEL OR HYDROGEN STORAGE

To understand the benefits and risks of using ammonia, it is necessary that some of its characteristics be considered. The first and most important is that it is a stable way of storing hydrogen and represents almost 18% of ammonia mass; its energy density per volume unit is 50% higher than that of liquid hydrogen but without all the dangerous hydrogen features. In addition, it is known that hydrogen storage has to be done at $-256°C$ if it is in the liquid phase. Ammonia is a fuel type that is free of carbon and GHG emissions with a high octane number and requires 30% less storage space than hydrogen.[103] The negligible inflammability under storage conditions makes ammonia safer than hydrogen in high amounts; furthermore, it does not weaken the storage tank metals like hydrogen. Its production and storage proceed through mature technologies that have been widely tested for decades (Figure 10.4). The catalyst employed in the Haber–Bosch process is an iron catalyst.[104] Based on reduced magnetite mineral (Fe_3O_4) enriched with aluminum or potassium oxides, this process demands much energy, for temperatures as high as $500°C$ can be reached.[105,106] However, there are studies on the application of catalysts based on ruthenium, nickel, and cobalt in order to carry out the process of obtaining ammonia at lower temperatures by electrically driven systems.[107,108] At present the Haber–Bosch process is one of the main GHG emitters, responsible for 1.2% of the global anthropogenic CO_2 emissions.[109,110] Last decade researchers tried to improve the electrochemical ammonia synthesis process, which will not have the negative effects of the classic Haber–Bosch process for obtaining ammonia; however, high activity and selectivity have not been achieved yet.[111–114]

Excellent reviews about power storage and engine fuels during the years 2022–2023 were published by researchers Wu,[103] Lee,[87] and Estevez.[115] High ignition temperature, high toxicity and corrosivity, in addition to possible high emissions of nitrogen oxide (NO_x), are the main disadvantages of ammonium and, as a consequence, it is not used as a monofuel for vehicles. Although the power developed by a NH_3 machine will never be equal to that of one with similar characteristics and working with fossil fuels, enthusiasts and small companies have proposed several projects to develop modified-internal-combustion-engine cars using pure ammonium or the air–ammonia–hydrogen mixture, which is described in detail in an article written by Portuguese researchers.[116]

Notwithstanding, by taking into account that ammonia is one of the best materials for hydrogen storage,[115] it can be used as e-fuel[87] through a route that includes the cracking process, purification of obtained hydrogen, and compression and liberation of hydrogen to fill the automobile tank (Figure 10.5).

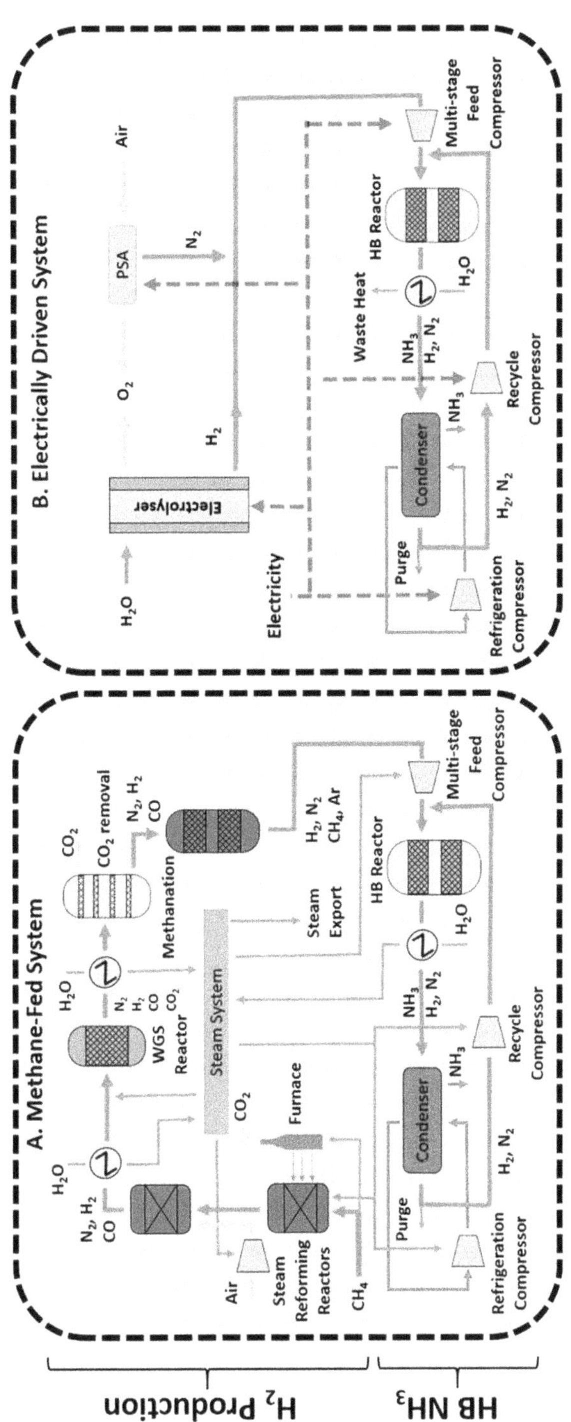

FIGURE 10.4 Ammonia production scheme by (a) the Haber–Bosch process and (b) an electrically powered alternative. Reproduced from [104] *Energy Environ. Sci.*, 2020,13, 331–344 with permission from the Royal Society of Chemistry.

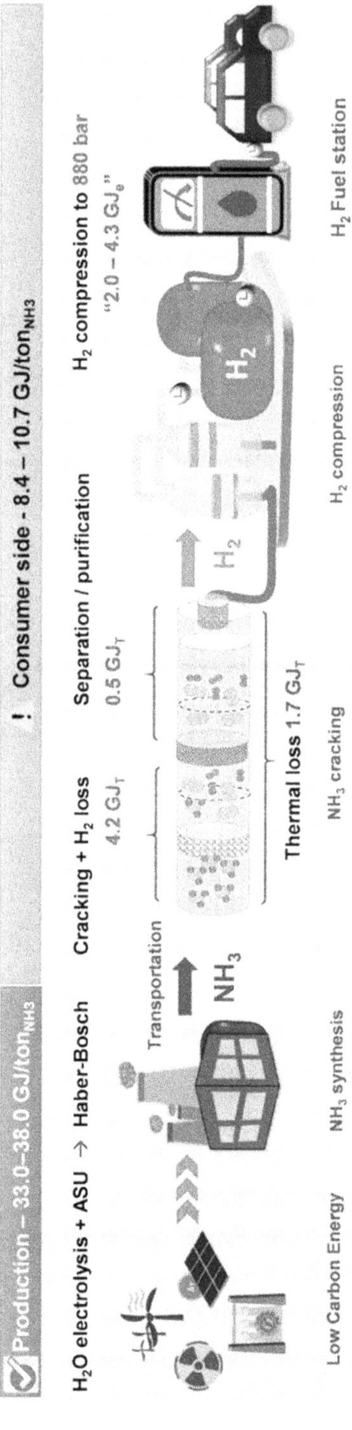

FIGURE 10.5 Power-to-power (PTP) energy consumption of NH$_3$ as a hydrogen energy carrier [117]. Reprinted (adapted) with permission from *ACS Energy Lett.* 2021, 6, 12, 4390–4394. https://doi.org/10.1021/acsenergylett.1c02189. Copyright 2021 American Chemical Society.

Due to the fact that this route is energetically highly demanding, the profit per produced energy is not evident.[117,118]

The energy consumption level throughout the production chain of e-fuels based on ammonia puts into question its "green" side for the environment despite the low carbon energy feature of this process.

10.6 MAGNESIUM FUEL POWER

Fuel cells produce energy and transform electrochemical energy into electricity through redox reactions between the anode and cathode in the electrochemical cell.[119,120] A natural source for generating this energy type is magnesium, which is the eighth most abundant metal in the Earth's crust and represents 2.5 wt.% of the Earth's resources. Magnesium sources are brucite [$Mg(OH)_2$], magnesite [$MgCO_3$], dolomite [$CaMg(CO_3)_2$], and seawater, where the presence of 1.8 billion tons is estimated; for this reason, it is an economic mineral on the market and safe to be manipulated. Due to its thermodynamic properties, it is an anodic material eligible for rechargeable batteries, for it provides high energy density and is highly inflammable as a powder with nitrogen and carbon dioxide; when it reacts with water, it releases 1,921 KJ/mol of energy without CO_2. Due to the foregoing characteristics, magnesium is an environmentally friendly energy source.[121]

Sobin and Abhilash (2013) reported studies on the profit of energy generated from the reaction between magnesium and water, which could be considered as an alternative fuel source; the engines that have been designed for this fuel type are known as Magnesium Power Engines. This engine is free of GHGs, for the reaction byproducts are magnesium oxide and water. In principle, this study considers that the engine power stems from the precursors in the engine with energy produced from the reaction between magnesium and water vapor. Liquid ammonia is used to absorb the thermal energy released by the reaction in the reaction chamber and an engine turbine is powered by hot ammonia gas.[122]

The German patent DE2360568A1 describes an internal combustion engine that uses magnesium vapor as fuel. The energy conversion is based on the fact that magnesium evaporates electrically and reacts with water vapor; the released hydrogen burns in the air and is employed in the compression engine under controlled conditions that are environmentally friendly.[123]

Diwan, M. et al. studied Mg/H_2O mixtures (Equation 10.1), which after an exothermic reaction generate hydrogen through self-sustainable combustion propagation under the effect of the metal particle size within an interval ranging from 75 to 150 μm, with a limited diffusion model for predicting the front velocity and thermal profile of the combustion wave, considering stoichiometric mixtures with an ignition temperature of 710°K (436°C) and a combustion experimental temperature of 2,050°K (1,776°C).[124]

$$M_g + 2H_2O = M_g(OH)_2 + H_2 \qquad (10.1)$$

The magnesium injection cycle engine is a system, where water inside the engine chamber reaches 500°C. Through this reaction with water, hydrogen is produced,

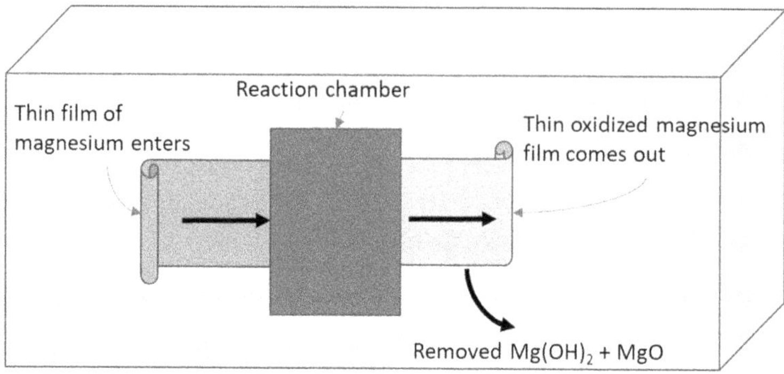

FIGURE 10.6   Magnesium fuel-cell battery (own elaboration).

which in turn reacts with oxygen to produce H_2O; these reactions generate the necessary power. However, magnesium hydroxide is produced as a product of the reaction.

In further studies at the Tokyo Technology Institute, this system was proposed with a magnesium renewable energy cycle and solar energy as the source of the pumped lasers for recycling magnesium used at Energy Genesis Cycle, Inc. The cell is a small fuel-cell type with a thin film inside the reaction chamber between the two electrodes and is constantly fed, thus increasing the generated energy (Figure 10.6); the reaction products, $Mg(OH)_2$ and MgO, are removed from the chamber and the Mg film is fed again. In previous chambers, the lack of removal of the oxidized film provoked low efficiency, which prompted this replacement to improve them; an automobile magnesium battery can have a weight of 200 kg.[126] Magnesium oxide is dissociated with laser irradiation efficiently.[125] The reduction of magnesium oxide (MgO) by means of a laser pumped by solar energy was reported by Yabe, T. et al., where a laser beam was focused on a mixture consisting of MgO and Si as the reducing agent, reaching a deposition efficiency of 2.3 mg/kJ;[126] the problematic aspect of magnesium is that its production is not clean as it is extracted by the electrolytic process and the high-temperature method.

As an emerging technology, magnesium can be an alternative to lithium; however, development is still needed to scale it up and massify it.

10.7   CONCLUSIONS

In 2021, approximately 25 million barrels of fossil-origin gasoline were consumed on a daily basis in the world. On the other hand, the production of e-fuels consumes much energy, and according to a study carried out by the Potsdam Institute for Climate Impact Research, the production volume expected for 2035 will not be enough to cover the demand. For this reason, in 2035, only 5 million out of 287 million cars (1.7%) in EU countries will be able to use synthetic e-fuel.[127] Furthermore, the price of e-fuels, produced from carbon dioxide using green hydrogen, is between 3 and 7 times higher than that of fossil fuels.[78] Coupling solar cells into electric vehicles does not solve the problem of using these vehicles for long distances; in addition, there are multiple problems with the production and use of lithium batteries. Magnesium fuel power is still in the research stage. The energy consumption and toxicity of ammonia do not allow for the massive application of this type of e-fuel. Hydrogen is the most promising e-fuel; however, it is still necessary to resolve the security issue and the distribution system and, in addition, expand green hydrogen production. For this reason, at least for the next two decades, most means of transportation will continue profiting from fossil-origin energy, mainly from gasoline.

REFERENCES

(1) Guéret, A.; Schill, W.-P.; Gaete-Morales, C. Not Flexible Enough? Impacts of Electric Carsharing on a Power Sector with Variable Renewables. arXiv February 29, 2024. https://arxiv.org/abs/2402.19380 (accessed 2024-04-08).

(2) Wehrmann, B. *EU delays crucial 2035 car emissions vote after German insistence on combustion engines.* https://www.cleanenergywire.org/news/eu-delays-crucial-2035-car-emissions-vote-after-german-insistence-combustion-engines

(3) Frost, R. EU 2035 Petrol and Diesel Car Ban: Germany Reaches Deal on Synthetic Fuels. *euronews.green.* 2023. https://twitter.com/Wissing/status/1639552963663261696?ref_src=twsrc%5Etfw%7Ctwcamp%5Etweetembed%7Ctwterm%5E1639552965563367425%7Ctwgr%5Efef00f35f79dd003ec49ae229504d074fb08e902%7Ctwcon%5Es2_&ref_url=https%3A%2F%2Fwww.euronews.com%2Fgreen%2F2023%2F03%2F22%2Feu-to-ban-petrol-and-diesel-cars-by-2035-heres-why-some-countries-are-pushing-back

(4) IRENA.org. *Kenya Spearheads Landmark Renewable Energy Initiative at Africa Climate Summit.* https://www.irena.org/News/pressreleases/2023/Sep/Kenya-Spearheads-Landmark-Renewable-Energy-Initiative-at-Africa-Climate-Summit

(5) Cariaga, C. *ThinkGeoEnergy's Top 10 Geothermal Countries 2023 – Power Generation Capacity.* https://www.thinkgeoenergy.com/thinkgeoenergys-top-10-geothermal-countries-2023-power-generation-capacity/.

(6) Shelare, S.; Kumar, R.; Gajbhiye, T.; Kanchan, S. Role of Geothermal Energy in Sustainable Water Desalination—A Review on Current Status, Parameters, and Challenges. *Energies* 2023, *16* (6). https://doi.org/10.3390/en16062901

(7) Ullah, A.; Altay Topcu, B.; Dogan, M.; Imran, M. Exploring the Nexus among Hydroelectric Power Generation, Financial Development, and Economic Growth: Evidence from the Largest 10 Hydroelectric Power-Generating Countries. *Energy Strategy Reviews* 2024, *52*, 101339. https://doi.org/10.1016/j.esr.2024.101339

(8) IRENA.org. *Wind Energy.* https://www.irena.org/Publications/2023/Nov/IRENA-EPO-Offshore-Wind-Energy-Patent-Insight-Report

(9) Feng, R. China's Energy Security and Geopolitical Imperatives: Implications for Formulating National Climate Policy. *Next Energy* 2024, *2*, 100034. https://doi.org/10.1016/j.nxener.2023.100034

(10) Yasmeen, R.; Zhang, X.; Sharif, A.; Shah, W. U. H.; Sorin Dincă, M. The Role of Wind Energy towards Sustainable Development in Top-16 Wind Energy Consumer Countries: Evidence from STIRPAT Model. *Gondwana Research* 2023, *121*, 56–71. https://doi.org/10.1016/j.gr.2023.02.024

(11) Lavrador, R. B.; Teles, B. A. D. S. Life Cycle Assessment of Battery Electric Vehicles and Internal Combustion Vehicles Using Sugarcane Ethanol in Brazil: A Critical Review. *Cleaner Energy Systems* 2022, *2*, 100008. https://doi.org/10.1016/j.cles.2022.100008

(12) Carvalho, E. N. D.; Pinho Brasil Júnior, A. C.; Brasil, A. C. D. M. Impact of Electric Vehicle Emissions in the Brazilian Scenario of Energy Transition and Use of Bioethanol. *Energy Reports* 2023, *10*, 2582–2596. https://doi.org/10.1016/j.egyr.2023.09.045

(13) Bonenkamp, T. B.; Middelburg, L. M.; Hosli, M. O.; Wolffenbuttel, R. F. From Bioethanol Containing Fuels towards a Fuel Economy That Includes Methanol Derived from Renewable Sources and the Impact on European Union Decision-Making on Transition Pathways. *Renewable and Sustainable Energy Reviews* 2020, *120*, 109667. https://doi.org/10.1016/j.rser.2019.109667

(14) Pulyaeva, V. N.; Kharitonova, N. A.; Kharitonova, E. N. Advantages and Disadvantages of the Production and Using of Liquid Biofuels. *IOP Conf. Ser.: Mater. Sci. Eng.* 2020, *976* (1), 012031. https://doi.org/10.1088/1757-899X/976/1/012031

(15) Okunev, V. S. Prospects for the Development of the Concept of a Safe Nuclear Reactor BREST of Maximum Limiting Power for Nuclear Energetics of the Middle of the XXI Century. *IOP Conf. Ser.: Mater. Sci. Eng.* 2021, *1100* (1), 012002. https://doi.org/10.1088/1757-899X/1100/1/012002

(16) Adamov, E. O.; Kaplienko, A. V.; Orlov, V. V.; Smirnov, V. S.; Lopatkin, A. V.; Lemekhov, V. V.; Moiseev, A. V. Brest Lead-Cooled Fast Reactor: From Concept to Technological Implementation. *Atomic Energy* 2021, *129* (4), 179–187. https://doi.org/10.1007/s10512-021-00731-w

(17) Byrne, L.; Bach, V.; Finkbeiner, M. Urban Transport Assessment of Emissions and Resource Demand of Climate Protection Scenarios. *Cleaner Environmental Systems* 2021, *2*, 100019. https://doi.org/10.1016/j.cesys.2021.100019

(18) Cordero, R. R.; Damiani, A.; Jorquera, J.; Sepúlveda, E.; Caballero, M.; Fernandez, S.; Feron, S.; Llanillo, P. J.; Carrasco, J.; Laroze, D.; Labbe, F. Ultraviolet Radiation in the Atacama Desert. *Antonie van Leeuwenhoek* 2018, *111* (8), 1301–1313. https://doi.org/10.1007/s10482-018-1075-z

(19) Rondanelli, R.; Molina, A.; Falvey, M. The Atacama Surface Solar Maximum. *Bulletin of the American Meteorological Society* 2015, *96* (3), 405–418. https://doi.org/10.1175/BAMS-D-13-00175.1

(20) Nooman AlMallahi, M.; Al Swailmeen, Y.; Ali Abdelkareem, M.; Ghani Olabi, A.; Elgendi, M. A Path to Sustainable Development Goals: A Case Study on the Thirteen Largest Photovoltaic Power Plants. *Energy Conversion and Management: X* 2024, *22*, 100553. https://doi.org/10.1016/j.ecmx.2024.100553

(21) Farag, M. M.; Bansal, R. C. Solar Energy Development in the GCC Region – a Review on Recent Progress and Opportunities. *International Journal of Modelling and Simulation* 2023, *43* (5), 579–599. https://doi.org/10.1080/02286203.2022.2105785

(22) Sukarso, A. P.; Kim, K. N. Cooling Effect on the Floating Solar PV: Performance and Economic Analysis on the Case of West Java Province in Indonesia. *Energies* 2020, *13* (9). https://doi.org/10.3390/en13092126

(23) Santoso, B.; Suyitno, S.; Wicaksono, F. *Design of 1 MWp Floating Solar Photovoltaic (FSPV) Power Plant in Indonesia*; AIP Publishing, 2019; Vol. 2097.

(24) Silalahi, D. F.; Gunawan, D. Solar Energy Potentials and Opportunity of Floating Solar PV in Indonesia. *Indonesia Post-Pandemic Outlook: Strategy towards Net-Zero Emissions by 2060 from the Renewables and Carbon-Neutral Energy Perspectives* 2022, 63–88.

(25) Singh, A. K.; Idrisi, A. H. Evolution of Renewable Energy in India: Wind and Solar. *Journal of The Institution of Engineers (India): Series C* 2020, *101* (2), 415–427.

(26) Escamilla-García, P. E.; Fernández-Rodríguez, E.; Jiménez-Castañeda, M. E.; Jiménez-González, C. O.; Morales-Castro, J. A. A Review of the Progress and Potential of Energy Generation from Renewable Sources in Latin America. *Latin American Research Review* 2023, *58* (2), 383–402. https://doi.org/10.1017/lar.2023.15

(27) Nevarez, D. An Interview with Oscar Ocampo: On Reaching Net Zero in Mexico. 2022.

(28) Blanco Corral, P. Proyecto Básico Planta Solar Fotovoltaica Puerto Peñasco (México). 2019.

(29) CDMX.gob. *Central Fotovoltaica de la Central de Abastos (CEDA)*. https://ciudadsolar. cdmx.gob.mx/programas/programa/central-fotovoltaica-de-la-central-de-abastos-ceda

(30) Aleksić, N.; Nikolić, D.; Šušteršič, V.; Jovanović, S. Development of the Modern Automotive Industry Based on the Solar Technology Application. *Mobility Vehicle Mech* 2022, *48* (3), 29–44.

(31) Heinrich, M.; Kutter, C.; Basler, F.; Mittag, M.; Alanis, L. E.; Eberlein, D.; Schmid, A.; Reise, C.; Kroyer, T.; Neuhaus, D. H. Peotential and Challenges of Vehicle Integrated Photovoltaics for Passenger Cars; 2020; Vol. 7, p 4229.

(32) *The beginning of true solar mobility — The moment Lightyear 0 went into production*. https://lightyear.one/articles/the-beginning-of-true-solar-mobility-the-moment-light year-0-went-into-production

(33) An, T. Study of a New Type of Electric Car: Solar-Powered Car. *IOP Conf. Ser.: Earth Environ. Sci.* 2021, *631* (1), 012118. https://doi.org/10.1088/1755-1315/631/1/012118

(34) Aptera Motors Corp. *Driven by the sun*. https://aptera.us

(35) Safdar, I.; Raza, A.; Khan, A. Z.; Ali, E. Feasibility, Emission and Fuel Requirement Analysis of Hybrid Car versus Solar Electric Car: A Comparative Study. *International Journal of Environmental Science and Technology* 2017, *14* (8), 1807–1818. https://doi. org/10.1007/s13762-017-1332-0

(36) T. Prasad; C. Nikhil; C. Sukrut; J. Suraj; M. Piyush. Solar-Powered Electric Vehicle Battery Charging: Integrating LT8490 for Sustainable and Efficient Energy Transfer in the Pursuit of a Green Transportation Future. In *2024 IEEE International Students' Conference on Electrical, Electronics and Computer Science (SCEECS)*; 2024; pp 1–6. https://doi.org/10.1109/SCEECS61402.2024.10482005

(37) Patil, C. D.; Gaikwad, P. P.; Puri, K. A.; Yadav, V. M.; Alase, A. B. Design and Implementation of Pv System for Electric Vehicle. *Tec Empresarial* 2024, *19* (1).

(38) Goetzberger, A.; Hebling, C.; Schock, H.-W. Photovoltaic Materials, History, Status and Outlook. *Materials Science and Engineering: R: Reports* 2003, *40* (1), 1–46. https://doi. org/10.1016/S0927-796X(02)00092-X

(39) Lacerda, J. S.; Van Den Bergh, J. C. Diversity in Solar Photovoltaic Energy: Implications for Innovation and Policy. *Renewable and Sustainable Energy Reviews* 2016, *54*, 331–340.

(40) Gangopadhyay, U.; Jana, S.; Das, S. State of Art of Solar Photovoltaic Technology. *Conference Papers in Energy* 2013, *2013*, 1–9. https://doi.org/10.1155/2013/764132

(41) Marques Lameirinhas, R. A.; Torres, J. P. N.; De Melo Cunha, J. P. A Photovoltaic Technology Review: History, Fundamentals and Applications. *Energies* 2022, *15* (5), 1823. https://doi.org/10.3390/en15051823

(42) Meria, F. H.; Algburi, S.; Ahmed, O. K. Impact of Porous Media on PV/Thermal System Performance: A Short Review. *Energy Reports* 2024, *11*, 1803–1819. https:// doi.org/10.1016/j.egyr.2024.01.046

(43) Ghosh, S.; Yadav, R. Future of Photovoltaic Technologies: A Comprehensive Review. *Sustainable Energy Technologies and Assessments* 2021, *47*, 101410. https://doi.org/ 10.1016/j.seta.2021.101410

(44) El Chaar, L.; El Zein, N. Review of Photovoltaic Technologies. *Renewable and Sustainable Energy Reviews* 2011, *15* (5), 2165–2175.

(45) Avrutin, V.; Izyumskaya, N.; Morkoç, H. Semiconductor Solar Cells: Recent Progress in Terrestrial Applications. *Superlattices and Microstructures* 2011, *49* (4), 337–364. https://doi.org/10.1016/j.spmi.2010.12.011

(46) Pastor, V. (2013, Enero 23). Plan Solar San Juan - Fabricación de Paneles Solares. Retrieved Abril 14, 2017, from Consultor de Telecomunicaciones y Electricidad: http:// Ingenieroandreotti.Blogspot.Mx/2013/01/Plan-Solar-San-Juan-Fabricacion-de.Html

(47) Sampaio, P. G. V.; González, M. O. A. Photovoltaic Solar Energy: Conceptual Framework. *Renewable and Sustainable Energy Reviews* 2017, *74*, 590–601. https://doi.org/10.1016/j.rser.2017.02.081

(48) Mundo-Hernández, J.; de Celis Alonso, B.; Hernández-Álvarez, J.; de Celis-Carrillo, B. An Overview of Solar Photovoltaic Energy in Mexico and Germany. *Renewable and Sustainable Energy Reviews* 2014, *31*, 639–649. https://doi.org/10.1016/j.rser.2013.12.029

(49) Powalla, M.; Paetel, S.; Ahlswede, E.; Wuerz, R.; Wessendorf, C. D.; Magorian Friedlmeier, T. Thin-film Solar Cells Exceeding 22% Solar Cell Efficiency: An Overview on CdTe-, Cu(In,Ga)Se2-, and Perovskite-Based Materials. *Applied Physics Reviews* 2018, *5* (4), 041602. https://doi.org/10.1063/1.5061809

(50) Ma, J.; Liu, Y.; Yao, Y.; Yang, X.; Meng, H.; Shi, J.; Meng, Q.; Liu, W. Suppressing Interface Recombination via Element Diffusion Regulation towards High-Efficiency Cd-Free Cu(In,Ga)Se$_2$ Solar Cells. *Nano Energy* 2024, *126*, 109641. https://doi.org/10.1016/j.nanoen.2024.109641

(51) Papež, N.; Dallaev, R.; Ţǎlu, Ş.; Kaštyl, J. Overview of the Current State of Gallium Arsenide-Based Solar Cells. *Materials* 2021, *14* (11). https://doi.org/10.3390/ma14113075

(52) Shahnooshi, F.; Orouji, A. A. Achieving High Photovoltaic Performance in Graphene/AlGaAs/GaAs Schottky Junction Solar Cells by Incorporating an InAlGaP Hole Reflector Layer. *Optics Communications* 2024, 130630. https://doi.org/10.1016/j.optcom.2024.130630

(53) Huang, Q.; Wang, Y.; Hu, X.; Yang, P.; Zhou, W.; Weng, G.; Akiyama, H.; Chu, J.; Chen, S. Effects of Localized Tensile Stress on GaAs Solar Cells Revealed by Absolute Electroluminescence Imaging and Distributed Circuit Modeling. *Solar Energy* 2024, *274*, 112541. https://doi.org/10.1016/j.solener.2024.112541

(54) Zheng, H.; Liu, Q.; Wang, Y.; Hu, J.; Zou, D.; Hou, S. Understanding and Weakening Photon Recycling in Solar Cells to Approach the Radiative Limit. *Advanced Materials* 2024, 2405063. https://doi.org/10.1002/adma.202405063

(55) Cheng, M.; Yang, X.; Li, S.; Wang, X.; Sun, L. Efficient Dye-Sensitized Solar Cells Based on an Iodine-Free Electrolyte Using l-Cysteine/l-Cystine as a Redox Couple. *Energy Environ. Sci.* 2012, *5* (4), 6290–6293. https://doi.org/10.1039/C1EE02540F

(56) Yoon, S.; Tak, S.; Kim, J.; Jun, Y.; Kang, K.; Park, J. Application of Transparent Dye-Sensitized Solar Cells to Building Integrated Photovoltaic Systems. *Building and Environment* 2011, *46* (10), 1899–1904. https://doi.org/10.1016/j.buildenv.2011.03.010

(57) Carey, G. H.; Abdelhady, A. L.; Ning, Z.; Thon, S. M.; Bakr, O. M.; Sargent, E. H. Colloidal Quantum Dot Solar Cells. *Chemical Reviews* 2015, *115* (23), 12732–12763. https://doi.org/10.1021/acs.chemrev.5b00063

(58) Emin, S.; Singh, S. P.; Han, L.; Satoh, N.; Islam, A. Colloidal Quantum Dot Solar Cells. *Solar Energy* 2011, *85* (6), 1264–1282. https://doi.org/10.1016/j.solener.2011.02.005

(59) Li, J.; Ni, J.; Guan, J.; Wang, R.; Wang, J.; Yang, Z.; Zhang, S.; Li, S.; Zhang, Y.; Li, J.; Cai, H.; Zhang, J. In Situ Passivation for High-Quality PbS Colloidal Quantum Dots Synthesized Using a PbBr2 Precursor. *Journal of Materials Science: Materials in Electronics* 2024, *35* (12), 827. https://doi.org/10.1007/s10854-024-12588-3

(60) Wang, Y.; Zhang, S.; Wang, J.; Ren, J.; Qiao, J.; Chen, Z.; Yu, Y.; Hao, X.-T.; Hou, J. Optimizing Phase Separation and Vertical Distribution via Molecular Design and Ternary Strategy for Organic Solar Cells with 19.5% Efficiency. *ACS Energy Lett.* 2024, 2420–2427. https://doi.org/10.1021/acsenergylett.4c00723

(61) Benanti, T. L.; Venkataraman, D. Organic Solar Cells: An Overview Focusing on Active Layer Morphology. *Photosynthesis Research* 2006, *87* (1), 73–81. https://doi.org/10.1007/s11120-005-6397-9

(62) Tyagi, V. V.; Rahim, N. A. A.; Rahim, N. A.; Selvaraj, J. A.L./Progress in Solar PV Technology: Research and Achievement. *Renewable and Sustainable Energy Reviews* 2013, *20*, 443–461. https://doi.org/10.1016/j.rser.2012.09.028

(63) Park, W.; Reo, Y.; Yang, W.; Choi, H.; Jeon, S.; Lim, B.; Liu, A.; Zhu, H.; Noh, Y.-Y. Two-Dimensional Dion-Jacobson Tin Perovskite Transistors with Enhanced Ambient Stability. *ACS Energy Letters* 2024, 2436–2445. https://doi.org/10.1021/acsenergylett.4c00620

(64) Park, D.-A.; Zhang, C.; Park, N.-G. Strain-Less Perovskite Film Engineered by Interfacial Molecule for Stable Perovskite Solar Cells. *ACS Energy Letters* 2024, 2428–2435. https://doi.org/10.1021/acsenergylett.4c00656

(65) Ibn-Mohammed, T.; Koh, S. C. L.; Reaney, I. M.; Acquaye, A.; Schileo, G.; Mustapha, K. B.; Greenough, R. Perovskite Solar Cells: An Integrated Hybrid Lifecycle Assessment and Review in Comparison with Other Photovoltaic Technologies. *Renewable and Sustainable Energy Reviews* 2017, *80*, 1321–1344. https://doi.org/10.1016/j.rser.2017.05.095

(66) Cocca, M.; Giordano, D.; Mellia, M.; Vassio, L. Free Floating Electric Car Sharing Design: Data Driven Optimisation. *Pervasive and Mobile Computing* 2019, *55*, 59–75.

(67) Sirojiddinova, I.; Aminboyev, A. S. Advantages and Disadvantages of Electric Vehicles and Hybrid Electric Vehicles. *American Journal of Engineering, Mechanics and Architecture (2993–2637)* 2023, *1* (10), 130–136.

(68) Ziegler, M. S.; Song, J.; Trancik, J. E. Determinants of Lithium-Ion Battery Technology Cost Decline. *Energy & Environmental Science* 2021, *14* (12), 6074–6098.

(69) Giglio, E. Extractivism and Its Socio-Environmental Impact in South America. Overview of the "Lithium Triangle". *América Crítica* 2021, *5* (1), 47–53.

(70) Sanchez-Lopez, M. D. Geopolitics of the Li-ion Battery Value Chain and the Lithium Triangle in South America. *Latin American Policy* 2023, *14* (1), 22–45.

(71) Luong, J. H. T.; Tran, C.; Ton-That, D. A Paradox over Electric Vehicles, Mining of Lithium for Car Batteries. *Energies* 2022, *15* (21). https://doi.org/10.3390/en15217997

(72) Mrozik, W.; Rajaeifar, M. A.; Heidrich, O.; Christensen, P. Environmental Impacts, Pollution Sources and Pathways of Spent Lithium-Ion Batteries. *Energy & Environmental Science* 2021, *14* (12), 6099–6121.

(73) Duan, X.; Zhu, W.; Ruan, Z.; Xie, M.; Chen, J.; Ren, X. Recycling of Lithium Batteries—a Review. *Energies* 2022, *15* (5), 1611.

(74) Badr, O.; Probert, S.; O'Callaghan, P. Atmospheric Methane: Its Contribution to Global Warming. *Applied Energy* 1991, *40* (4), 273–313.

(75) Atsbha, T. A.; Yoon, T.; Seongho, P.; Lee, C.-J. A Review on the Catalytic Conversion of CO_2 Using H_2 for Synthesis of CO, Methanol, and Hydrocarbons. *Journal of CO2 Utilization* 2021, *44*, 101413. https://doi.org/10.1016/j.jcou.2020.101413

(76) Dahmen, N.; Arnold, U.; Djordjevic, N.; Henrich, T.; Kolb, T.; Leibold, H.; Sauer, J. High Pressure in Synthetic Fuels Production. *The Journal of Supercritical Fluids* 2015, *96*, 124–132. https://doi.org/10.1016/j.supflu.2014.09.031

(77) Teimouri, Z.; Abatzoglou, N.; Dalai, A. K. Kinetics and Selectivity Study of Fischer–Tropsch Synthesis to C5+ Hydrocarbons: A Review. *Catalysts* 2021, *11* (3). https://doi.org/10.3390/catal11030330

(78) Cames, M.; Chaudry, S.; Göckeler, K.; Kasten, P.; Kurth, S. *E-Fuels versus DACCS*; E-fuel versus greenhouse gas removals in aviation; Oko-Institute, 2021; p 39. tps://www.transportenvironment.org/wp-content/uploads/2021/08/2021_08_TE_study_efuels_DACCS.pdf

(79) Singh, H.; Li, C.; Cheng, P.; Wang, X.; Liu, Q. A Critical Review of Technologies, Costs, and Projects for Production of Carbon-Neutral Liquid e-Fuels from Hydrogen and Captured CO 2. *Energy Advances* 2022, *1* (9), 580–605.

(80) Saint Petersburg Mining University; Kuznetsova, E. A.; Cherepovitsyna, A. A.; Saint Petersburg Mining University. Carbon dioxide utilization and circular economy: The world, Russia and the Arctic. СИР 2021, *24* (4–2021), 42–55. https://doi.org/10.37614/2220-802X.4.2021.74.004

(81) Bartlett, J. Chile Apuesta al Hidrogeno Verde. *Finanzas & Desarrollo.* New York 2022, 42–43.

(82) Kitman, J. Faux Fuel: Can Synthetic Gasoline Save Internal Combustion? *Car and Driver* 2021, *67* (3), 14–15.

(83) Stan, C. Batteriebetriebene Elektroautos Und Verbrennungsmotoren Mit Klimagerechten Kraftstoffen: Ergänzen Statt Ersetzen. In Klimagerechte *Energieszenarien der Zukunft: Mobilität, Heizung, Industrie-mehr als nur Elektroautos, luftsaugende Wärmepumpen und Windräderstrom*; Springer, 2024; pp 19–41.

(84) Romo Martínez, A. C. The Environmental Governance System around the Implementation of Sustainable Aviation Fuels in Norway: A Case Study on e-Fuels and Carbon Capture and Utilization Value Chain. *Norwegian University of Life Science* 2021.

(85) García-Bordejé, E.; González-Olmos, R. Advances in Process Intensification of Direct Air CO2 Capture with Chemical Conversion. *Progress in Energy and Combustion Science* 2024, *100*, 101132.

(86) Hydrogen Council, McKinsey & Company. *Global Hydrogen Flows: Hydrogen Trade as a Key Enabler for Efficient Decarbonization*; 2022. https://hydrogencouncil.com/wp-content/uploads/2022/10/Global-Hydrogen-Flows.pdf

(87) Lee, B.; Winter, L. R.; Lee, H.; Lim, D.; Lim, H.; Elimelech, M. Pathways to a Green Ammonia Future. *ACS Energy Letters* 2022, *7* (9), 3032–3038. https://doi.org/10.1021/acsenergylett.2c01615

(88) IEA. *Global Hydrogen Review 2023*; International Energy Agency, 2023.

(89) Sanoja-López, K. A.; Loor-Molina, N. S.; Luque, R. An Overview of Photocatalyst Eco-Design and Development for Green Hydrogen Production. *Catalysis Communications* 2024, 106859. https://doi.org/10.1016/j.catcom.2024.106859

(90) Ogden, J. M. PROSPECTS FOR BUILDING A HYDROGEN ENERGY INFRA-STRUCTURE. *Annu. Rev. Energy. Environ.* 1999, *24* (1), 227–279. https://doi.org/10.1146/annurev.energy.24.1.227

(91) Hassan, Q.; Azzawi, I. D. J.; Sameen, A. Z.; Salman, H. M. Hydrogen Fuel Cell Vehicles: Opportunities and Challenges. *Sustainability* 2023, *15* (15), 11501. https://doi.org/10.3390/su151511501

(92) Leal, V.; Villasuso, M.; Sulmont, A.; Martínez, C. *El Potencial Industrial de México En Las Cadenas de Valor Del Hidrógeno Verde*; Programa de las Naciones Unidas para el Desarrollo, 2023. https://www.undp.org/es/mexico/publicaciones/el-potencial-industrial-de-mexico-en-las-cadenas-de-valor-del-hidrogeno-verde

(93) Reddi, K.; Elgowainy, A.; Rustagi, N.; Gupta, E. Impact of Hydrogen Refueling Configurations and Market Parameters on the Refueling Cost of Hydrogen. *International Journal of Hydrogen Energy* 2017, *42* (34), 21855–21865. https://doi.org/10.1016/j.ijhydene.2017.05.122

(94) Haghani, M.; Sprei, F.; Kazemzadeh, K.; Shahhoseini, Z.; Aghaei, J. Trends in Electric Vehicles Research. *Transportation Research Part D: Transport and Environment* 2023, *123*, 103881. https://doi.org/10.1016/j.trd.2023.103881

(95) Sandaka, B. P.; Kumar, J. Alternative Vehicular Fuels for Environmental Decarbonization: A Critical Review of Challenges in Using Electricity, Hydrogen, and Biofuels as a Sustainable Vehicular Fuel. *Chemical Engineering Journal Advances* 2023, *14*, 100442. https://doi.org/10.1016/j.ceja.2022.100442

(96) Le, T. T.; Sharma, P.; Bora, B. J.; Tran, V. D.; Truong, T. H.; Le, H. C.; Nguyen, P. Q. P. Fueling the Future: A Comprehensive Review of Hydrogen Energy Systems and Their Challenges. *International Journal of Hydrogen Energy* 2024, *54*, 791–816. https://doi.org/10.1016/j.ijhydene.2023.08.044

(97) Fereidounizadeh, N. Hydrogen Fuel in Sweden, a Comparative Study of Five Countries. 2021.

(98) US Department of Energy. Alternative Fuels Data Center, 2024. https://afdc.energy.gov/vehicles/how-do-fuel-cell-electric-cars-work

(99) Wallner, T.; Miers, S. A. Internal Combustion Engines, Alternative Fuels For. In *Electric, Hybrid, and Fuel Cell Vehicles*; Springer, 2021; pp 27–66.

(100) Wróbel, K.; Wróbel, J.; Tokarz, W.; Lach, J.; Podsadni, K.; Czerwiński, A. Hydrogen Internal Combustion Engine Vehicles: A Review. *Energies* 2022, *15* (23), 8937.

(101) Yoshida, T.; Kojima, K. Toyota MIRAI Fuel Cell Vehicle and Progress Toward a Future Hydrogen Society. *Interface Magazine* 2015, *24* (2), 45–49. https://doi.org/10.1149/2.F03152if

(102) Reuters. *Japan to Collaborate with US on Cutting Floating Offshore Wind Costs*; 2024. https://www.reuters.com/business/energy/

(103) Wu, S.; Salmon, N.; Li, M. M.-J.; Bañares-Alcántara, R.; Tsang, S. C. E. Energy Decarbonization via Green H$_2$ or NH$_3$? *ACS Energy Letters* 2022, *7* (3), 1021–1033. https://doi.org/10.1021/acsenergylett.1c02816

(104) Smith, C.; Hill, A. K.; Torrente-Murciano, L. Current and Future Role of Haber–Bosch Ammonia in a Carbon-Free Energy Landscape. *Energy & Environmental Science* 2020, *13* (2), 331–344. https://doi.org/10.1039/C9EE02873K

(105) Leigh, G. J. Haber-Bosch and Other Industrial Processes. In *Catalysts for Nitrogen Fixation: Nitrogenases, Relevant Chemical Models and Commercial Processes*; Smith, B. E., Richards, R. L., Newton, W. E., Eds.; Springer Netherlands: Dordrecht, 2004; pp 33–54. https://doi.org/10.1007/978-1-4020-3611-8_2

(106) Ribeiro, D.; Faculdade de Ciências da Universidade do Porto. Processo de Haber-Bosch. *Rev. Ciência Elem.* 2013, *1* (1). https://doi.org/10.24927/rce2013.031

(107) Humphreys, J.; Lan, R.; Tao, S. Development and Recent Progress on Ammonia Synthesis Catalysts for Haber–Bosch Process. *Adv Energy and Sustain Res* 2021, *2* (1), 2000043. https://doi.org/10.1002/aesr.202000043

(108) Rouwenhorst, K. H. R.; Van der Ham, A. G. J.; Lefferts, L. Beyond Haber-Bosch: The Renaissance of the Claude Process. *International Journal of Hydrogen Energy* 2021, *46* (41), 21566–21579. https://doi.org/10.1016/j.ijhydene.2021.04.014

(109) Nørskov, J.; Chen, J.; Miranda, R.; Fitzsimmons, T.; Stack, R. *Sustainable Ammonia Synthesis – Exploring the Scientific Challenges Associated with Discovering Alternative, Sustainable Processes for Ammonia Production*; United States, 2016. https://doi.org/10.2172/1283146

(110) Wang, M.; Khan, M. A.; Mohsin, I.; Wicks, J.; Ip, A. H.; Sumon, K. Z.; Dinh, C.-T.; Sargent, E. H.; Gates, I. D.; Kibria, M. G. Can Sustainable Ammonia Synthesis Pathways Compete with Fossil-Fuel Based Haber–Bosch Processes? *Energy & Environmental Science* 2021, *14* (5), 2535–2548. https://doi.org/10.1039/D0EE03808C

(111) Hatzell, M. C. A Decade of Electrochemical Ammonia Synthesis. *ACS Energy Letters* 2022, *7* (11), 4132–4133. https://doi.org/10.1021/acsenergylett.2c02335

(112) MacLaughlin, C. Role for Standardization in Electrocatalytic Ammonia Synthesis: A Conversation with Leo Liu, Lauren Greenlee, and Douglas MacFarlane. *ACS Energy Letters* 2019, *4* (6), 1432–1436. https://doi.org/10.1021/acsenergylett.9b01123

(113) Capdevila-Cortada, M. Electrifying the Haber–Bosch. *Nature Catalysis* 2019, *2* (12), 1055. https://doi.org/10.1038/s41929-019-0414-4

(114) Soloveichik, G. Electrochemical Synthesis of Ammonia as a Potential Alternative to the Haber–Bosch Process. *Nature Catalysis* 2019, *2* (5), 377–380. https://doi.org/10.1038/s41929-019-0280-0

(115) Estevez, R.; López-Tenllado, F. J.; Aguado-Deblas, L.; Bautista, F. M.; Romero, A. A.; Luna, D. Current Research on Green Ammonia (NH$_3$) as a Potential Vector Energy for Power Storage and Engine Fuels: A Review. *Energies* 2023, *16* (14), 5451. https://doi.org/10.3390/en16145451

(116) Cardoso, J. S.; Silva, V.; Rocha, R. C.; Hall, M. J.; Costa, M.; Eusébio, D. Ammonia as an Energy Vector: Current and Future Prospects for Low-Carbon Fuel Applications in Internal Combustion Engines. *Journal of Cleaner Production* 2021, *296*, 126562. https://doi.org/10.1016/j.jclepro.2021.126562

(117) Chatterjee, S.; Parsapur, R. K.; Huang, K.-W. Limitations of Ammonia as a Hydrogen Energy Carrier for the Transportation Sector. *ACS Energy Letters* 2021, *6* (12), 4390–4394. https://doi.org/10.1021/acsenergylett.1c02189

(118) Lazouski, N.; Limaye, A.; Bose, A.; Gala, M. L.; Manthiram, K.; Mallapragada, D. S. Cost and Performance Targets for Fully Electrochemical Ammonia Production under Flexible Operation. *ACS Energy Letters* 2022, *7* (8), 2627–2633. https://doi.org/10.1021/acsenergylett.2c01197

(119) Olabi, A. G.; Wilberforce, T.; Abdelkareem, M. A. Fuel Cell Application in the Automotive Industry and Future Perspective. *Energy* 2021, *214*, 118955. https://doi.org/10.1016/j.energy.2020.118955

(120) Salminen, J.; Steingart, D.; Kallio, T. Chapter 15 - Fuel Cells and Batteries. In *Future Energy*; Letcher, T. M., Ed.; Elsevier: Oxford, 2008; pp 259–276. https://doi.org/10.1016/B978-0-08-054808-1.00015-6

(121) Aurbach, D.; Lu, Z.; Schechter, A.; Gofer, Y.; Gizbar, H.; Turgeman, R.; Cohen, Y.; Moshkovich, M.; Levi, E. Prototype Systems for Rechargeable Magnesium Batteries. *Nature* 2000, *407* (6805), 724–727. https://doi.org/10.1038/35037553

(122) Santhosh, S.; Nair, A. Magnesium and the Magnesium Power Engine.

(123) Studenski, U. Internal Combustion Engine Using Magnesium Vapour as Fuel - by Reacting It with Water Vapour and Burning Liberated Hydrogen in Air. DE2360568A1, 1973.

(124) Diwan, M.; Hanna, D.; Shafirovich, E.; Varma, A. Combustion Wave Propagation in Magnesium/Water Mixtures: Experiments and Model. *Chemical Engineering Science* 2010, *65* (1), 80–87.

(125) Sato, Y.; Yabe, T.; Sakurai, Y.; Mohamed, M.; Uchida, S.; Baasandash, C.; Ohkubo, T.; Mori, Y.; Sato, H. Experimental Study of Magnesium Production with Laser for Clean Energy Cycle; *American Institute of Physics*, 2008; Vol. 997, pp 546–552.

(126) Yabe, T.; Suzuki, Y.; Satoh, Y. Renewable Energy Cycle with Magnesium and Solar-Energy-Pumped Lasers. *REPQJ* 2014, 76–80. https://doi.org/10.24084/repqj12.236

(127) Ueckerdt, F.; Odenweller, A. E-Fuels - Aktueller Stand Und Projektionen, 2023. https://www.pik-potsdam.de/members/Ueckerdt/E-Fuels_Stand-und-Projektionen_PIK-Potsdam.pdf

Acknowledgments

We acknowledge first all the contributors who made possible this book. We acknowledge CRC Editorial Group for their positive reception of this publication. We also thank *Consejo Nacional de Humanidades, Ciencia y Tecnología* (CONAHCyT) (Project Cf-191973) and the Mexican Petroleum Institute (Project Y.62011) for their financial support and the opportunity granted to get this book done.

Index

Pages in *italics* refer to figures and pages in **bold** refer to tables.

A

acetates, 98–99
acid-catalyzed reaction, 162
acidity, 158, 164
Adani Green Energy Limited, 173
Ag, 141
alcohols, 95–103
Al Dhafra, 173
aldehydes, 98–101
algae, 108, 110–111, 116, *117*
AlkyClean process, 81
alkylate, 69, *70*, 70–82
alkylation, 34, 53, 60, *70–81*, 81–82, **82**
    first commercial alkylation plant, 71
    reaction, 71–74, *75*, 75–77, *78*, 78–80
    strong acids, 72, 75, 82
alumina, 55–56, **56**, 63
ammonia as a hydrogen energy carrier, *182*
ammonia as e-fuel or hydrogen storage, 180–183
ammonia production, *181*
antiknock, 88–90, 92–93, 99, 101–103
aromatic, 33–37, 39–44
Au, 141, **145**
automotive, 1, 9
    market, 70
aviation fuel, 124–126
    hydrocarbons, 126, *132*

B

battery, 9
bifunctional catalysts
    acidity, 158, 160
    β zeolite, 158
    cordierite monoliths, 159
    polyoxometalates, 160
    SAPO-11, 160
    SBA-15, 160
    zeolites, 156, 158, 160
    ZSM-5, 158–159, 164
biodiesel, 5
biodiversity, 2, 8
bioethanol, 172
biofuel, **4**, 5, 7, 10
biomass, 2, **4**, 5, 7, 108, 110–114, 116, 118–122
bio-syngas, 113
BREST-OD-300 reactor, 172
butane isomerization plant, 52
    butanol, 95, *95*, **96**, 97, 100–102

C

CAMERE process, *136*
carbenium, 75, 77
    mechanism, 74, *75*, 75, 77, *78*
carbon dioxide (CO_2), 98, 124, 126–154,
        156–157, 172, 176, 183–184
carbon emissions, 71
carbon monoxide (CO), 89, 91–93, 97–101
carbon recycling international, *137*
carbonates, 92, 99, *99*
catalyst, **69**, 71–72, 75–77, 79–82, 124, 126–129,
        131, 133–137, *140*, 141–144
    adsorption, 126–127, 129, 131, 133,
        140–141
    aging, 25
    Brönsted acids, 77, 82
    Co, Fe, In, **128**, 128–131, *132*, 133–135,
        137–139, 142, **144**
    CO reforming, 125
    copper (Cu), 133, *140*, 141
    deactivation, 33, 39, 41, 45
    hydrofluoric acid, 71
    for the hydroisomerization, 55, 62
    performance, 44, 45
    Lewis acids, 71, 73, 77–80, 82
    poisoning, 24
    sulfuric acid, 71, 72, 74, 76
catalytic, 1, 5, 8
    disintegration, 14–15
    hydroprocessing, 114–116
    support, 38
Central de abastos photovoltaic plant, 173
chain growth, 160
chemical degradation, 24
chloroaluminates, 77–79
clean energy, 171
clean fuels, 154, 156
CO hydrogenation, 154–155
coal gasification, 155
coke formation, 18, 20, 25
combustion, 1–2, 5, 7–8, 87–93, 95, 97–103
composition of oil, 13
cooperative fuel research (CFR), 88–89
cracking, 14, 16, 18–25, 27
cracking reactions, 20–21, 24–26
crafting naphtha reforming, 42
crude oil, 2, 7–8, 10, 69, 80
cyclopentadienyl manganese tricarbonyl (CMT),
    90

D

E

F

G

H

I

For Product Safety Concerns and Information please contact our EU
representative GPSR@taylorandfrancis.com
Taylor & Francis Verlag GmbH, Kaufingerstraße 24, 80331 München, Germany

www.ingramcontent.com/pod-product-compliance
Ingram Content Group UK Ltd.
Pitfield, Milton Keynes, MK11 3LW, UK
UKHW022315100726
473146UK00009B/476